CLASS, NATION AND

T0256590

Anthropology, Culture and Society

Series Editors:
Professor Thomas Hylland Eriksen, University of Oslo
Dr Katy Gardner, University of Sussex
Dr Jon P. Mitchell, University of Sussex

CLASS, NATION AND IDENTITY

The Anthropology of Political Movements

JEFF PRATT

Pluto Press
LONDON

First published 2003
by PLUTO PRESS
345 Archway Road, London N6 5AA

www.plutobooks.com

British Library Cataloguing in Publication Data
A catalogue record for this book is available from
the British Library

ISBN 978 0 7453 1671 0 paperback

Library of Congress Cataloging in Publication Data
Pratt, Jeff C.
    Class, nation and identity / Jeff Pratt.
        p. cm. — (Anthropology, culture, and society)
    Includes bibliographical references and index.
        ISBN 0–7453–1672–7 (hc) — ISBN 0–7453–1671–9 (pb)
    1. Nationalism. 2. Social classes—Political aspects.
    3. Nationalism—Europe—Case studies. 4. Europe—Social conditions.
    I. Title. II. Series.
    JC311 .P683 2003
    320.54'094—dc21
                                                                    2002010407

10   9   8   7   6   5   4   3   2   1

Designed and produced for Pluto Press by
Chase Publishing Services, Fortescue, Sidmouth EX10 9QG
Typeset from disk by Stanford DTP Services, Towcester
Printed on demand in the EU by
CPI Antony Rowe, Chippenham and Eastbourne, England

CONTENTS

ACKNOWLEDGEMENTS

The first stimulus for this book came out of the experience of research in Central Italy in the early 1970s. This was an environment structured by the pervasive conflict between Catholic and communist political movements, and it was here that I discovered that as a fieldworker my own political identity was more important than my nationality in most social contexts. It was also here that I encountered the contrast between rural and urban social worlds, and began to hear the many ways that these are represented. This contrast has profound importance in European political history and constitutes a recurring theme in this study.

A second and in some ways more enduring stimulus for this book comes from teaching, from the wish to make information about political movements and their interpretation more accessible. The book is the result of long collaboration and challenge from colleagues and students at Sussex University on courses in political anthropology and contemporary European studies. The arguments have been modified and reformulated in the face of their alternative perspectives and constant scepticism. I would particularly like to acknowledge my indebtedness to Zdenek Kavan, with whom I have taught a Masters course on nationalism, ethnicity and citizenship for many years. I have learned much from him about European history and political theory, and we have both had to clarify our arguments in the face of our students' steadfast refusal to be convinced. Work on Yugoslavia is notoriously risky for the non-specialist, and I would not have embarked on the case study which became chapter 7 without the presence at the Sussex European Institute in the mid-1990s of a very substantial group of scholars, including Cornelia Sorabji, Ivan Vejvoda and Mary Kaldor.

The book was researched and written during sabbatical leave granted by the Sussex anthropology department, and the text was edited with painstaking care by Jenny Money. Tom Wilson, Peter Luetschford and Jon Mitchell gave me detailed comments on the manuscript, and the latter as series editor for Pluto gave me much support in the stages of preparation. To all of them my profound thanks.

All books have some loose ends: ideas not worked through, sources and suggestions not followed up. Personal circumstances mean that there is more of this unfinished business than I would have liked, but it was time to stop. It is dedicated to the memory of Assuntina Angeli, *compagna della vita.*

1 INTRODUCTION

In 1873 the socialist workers of Breslavia (Wroclaw), which was then in Germany, ten years after the death of the pioneer workers' leader Lassalle, dedicated a new red flag. On the front it bore ... an inscription to ... Lassalle, surmounted by the motto Liberty, Equality, Fraternity ... During Bismarck's antisocialist law the flag was smuggled into Switzerland. Under Hitler, between 1933 and 1945, it was carefully kept, first buried in an allotment garden, later in the cellar of a plumber, who refused to give it up to the Red Army officers who came to salute it in 1945. When Breslavia became Polish and was renamed Wroclaw, the keeper of the flag transported it to West Germany to hand it over to the Social-Democratic Party, which, presumably, still has it ...

(Hobsbawm: 1984b: 67)

The increasing prominence of ethnic and national movements at a global level over the last 20 years has led to a phenomenal growth in research, and the resulting studies have generated a sense of intellectual excitement and challenge which has permeated a range of intellectual disciplines, from literary studies to international relations. Two books, both published in 1983, have achieved canonical status. Gellner's *Nations and Nationalism* opened up the relationship between nationalist movements and modernity, the question of historical continuity in group identity, and the relative salience of different cultural elements in group definition. Benedict Anderson's *Imagined Communities* directed attention to the importance of experience, memory and imagination, the emotional power of these movements and their actual or potential recourse to violence. This is a rich research agenda, but also significantly different from the one that has dominated research on other kinds of politics. Why should this be? And where are the distinctive features of ethnic or national politics located?

The inevitable comparison here is with the analysis of class politics. In the first part of the twentieth century it had been class rather than

1

ethnicity which had acted as the purported motor of European political history, and had been the dominant subject of intellectual inquiry. The celebrated death of class as a 'master narrative' has often in practice meant the death of class as any kind of narrative, and this makes it harder to get the issues back into perspective. There is growing amnesia about both the range of class movements and their political scope; they are remembered chiefly as struggles around economic interests, generated in the workplace, in a now largely superseded system of industrial production. Their death, for some analysts (Escobar 1992; Touraine 1985), allowed for the birth of 'new social movements', the release of energies into more complex and differentiated forms of political struggle, particularly around questions of identity.

Yet, as the best studies have shown, class politics was always more complex than that. Even the short quotation from Hobsbawm above is a reminder of the bitter divisions within class politics, the jealous guarding of symbols, and political allegiances which were more enduring than national boundaries. The kinds of explanation and interpretation which emerged as paradigmatic in the study of ethnicity and nationalism can be, and have been, deployed in the study of class. In this terrain we are also faced with all the dimensions of modernity, with the dislocation and concentration of economic and cultural capitals, with the mass movements of people and rapid cultural evolution. Here too we find complex historical narratives which shape class identity and future aspirations, we find 'imagined communities' and the transmission of memories. Here too we encounter the same interpretative issues about how identity is 'constructed', and how specific experiences become class experiences. Above all, we also find passion. It is astonishing that the point has to be made, but such is the mesmeric hold of the canonical texts that it is worth repeating: passion is not an ingredient unique to the politics of nationalism.

The first purpose of this book is to re-examine the ways in which class and ethnic or nationalist politics have been analysed, and to open up issues which are of general importance in political anthropology. In suggesting that the wide-ranging research agenda which has illuminated our understanding of nationalism can also be applied to class movements, and vice versa, I am not suggesting that these movements are indistinguishable. The problem lies elsewhere, in the use of different paradigms for the study of political process. These have made the two kinds of politics incommensurable, in terms of the type of person who acts, and the kind of world they inhabit. Put simply, in one world we find economic categories of people driven by material interests, in the other cultural subjects consumed by passions.

This introductory chapter will start by exploring some of the theoretical texts which have contributed to this view of a qualitative difference between the politics of nationalism and all other struggles. Out of it will emerge a more open and less reified view of what 'economics' or 'culture' is about, and suggestions about how we can develop a common approach to political movements. The next sections examine some of the key issues in the analysis of nationalist and class politics, and explore the implications of adopting a more unified approach, one which keeps the richness of insight which can be found in both fields of study. Demonstrating the value of this approach cannot be done through comparative surveys or generalising overviews: it requires a depth of ethnographic and historical information which reveals the interaction of multiple processes. For that reason I decided to build up the argument through case studies, all taken from the southern parts of Europe, and all dealing with large-scale political action which made a sustained impact on society and the state. The first group of case studies deals with class-based mobilisation, the second with nationalism, followed by shorter examples of movements which combine elements of both. The cases are grouped together, but each chapter covers similar ground, examining the broad historical context within which these movements emerged, the construction of collective identities, and their characteristic forms of organisation and political action. In intermediate sections and in the conclusion I will draw out the comparative dimensions and explore their relevance for theoretical work in this field.

TWO CLASSIC TEXTS AND THEIR JOKES

Benedict Anderson's *Imagined Communities* (1983) opens up stimulating questions about the changing representations of time in political action, and on the role of history, memory and the imagination in the formation of identity. The focus is also on political passion: in numerous reviews of the book (starting with Kitching 1985), Anderson's superiority to Gellner is seen in his ability to explain the passion of nationalism, and it is linked to the claim that nationalism plays much the same part in the life of people as old-time religion. All this is important and extremely valuable, and it is illustrated by a much-quoted and rather bizarre joke. In commenting on the religion-like qualities of nationalism, he remarks that most states instituted a tomb of the unknown warrior, but tombs of fallen liberals or unknown Marxists are rather rare. Anderson knew very well that at the time he was writing, the world was full of much-venerated tombs and statues to known Marxists, even if many have since been exhumed or their statues melted down. You cannot have unknown

Marxists, if by that is intended writers, but you can and did have many commemorations of anonymous workers, who had been martyrs or heroes in the cause of the revolution. There was also a whole Cold-War industry, much of it located in Princeton and Harvard (Shore 1990: 59), saying that communism was just like old-time religion, full of fanatics and prophets, whose faith was immune to argument and who exploited the irrational masses.

Does the joke matter? I think it does. It is part of a series of rhetorical moves which digs a ditch around the study of nationalism, and ends up locking the study of class and ethnic politics into incommensurable paradigms. Both class and ethnicity are shorthand terms for a variety of political movements; obviously there are distinctions to be made, but they will not emerge if we make the assumption that ethnic politics is in any sense unique because it has a cultural dimension, or because it involves identity construction, or because it generates political passion or violence. If we do not break with this paradigm, we underwrite the inevitability of ethnic politics.

Ernest Gellner's book *Nations and Nationalism* (1983) also has a famous joke, the claim that Marxists, because they were unable to explain the failure of revolutions and the power of national solidarity, had a 'wrong address theory' of nationalism: 'The awakening message was intended for classes, but by some terrible postal error was delivered to nations.' This is one of four false theories of nationalism which are given short shrift at the end of his study – 'not one of these theories is remotely tenable' (Gellner 1983: 130). Again, it is not clear whether this joke was produced in good faith, and its popularity is strange, since it acts as a smokescreen for the argument.

The brevity of the reference should not mislead us, since much of the book is concerned with debunking Marxist theories of conflict and the polarisation of class relations. 'Capital, like capitalism,' says Gellner (1983: 97) 'seems an overrated category'. Instead he writes about industrial society, where he claims, 'Stratification and inequality do exist, and sometimes in extreme form; nevertheless they have a muted and discrete quality, attenuated by a kind of gradualness of the distinctions of wealth and standing' (1983: 25). I will come back to the extreme forms of inequality and dichotomous views of class, after looking at what Gellner identifies as the real locus of conflict in industrial society.

Nationalism first emerges in the transition to industrialism, a period of turbulent readjustment when political and cultural boundaries become more congruent. Nationalism is the process whereby a culture is endowed with a political roof. Two issues arise: first, what is a culture?; and secondly, under what circumstances is the transition turbulent or violent? Gellner admits that culture is an elusive concept, but that in general he is using it in an anthropological sense, to distinguish it from *Kultur*, or high culture. This does not clarify a great deal, because Gellner's usage is not just variable, but actually builds a key part of the argument on a slippage between two kinds of referent. The dominant meaning is associated with language, and he discusses the enormous gap between the number of languages in the world and the number of actual or potential nationalisms. This is illustrated through a celebrated and witty account of the alienation of a cultural minority, the Ruritanian migrants, who find themselves in Megalomania. In a later article he returns to the same theme. 'The modern world has produced societies in which the division of labour is very advanced, the occupational structure is highly unstable and most work is semantic and communicative rather than physical' (Gellner 1997: 85). In that world

individuals find themselves in very stressful situations unless the nationalist requirement of congruence between a man's culture and that of his environment is satisfied. Without such a congruence, life is hell. Hence the deep passion which, according to Perry Anderson, is absent from my theory. (Gellner 1997: 84)

In these phrases about semantics and communication, the ground has been prepared for a different use of the term culture, one which appears further on in the 1983 monograph. On the issue of whether industrialism will continue to generate conflict, Gellner says,

We have yet to discuss the difficult and important question whether advanced industrialism, as such, in any case constitutes a shared culture, overruling the – by now irrelevant – differences of linguistic idiom. When men [*sic*] have the same concepts, more or less, perhaps it no longer matters whether they use different words to express them. (Gellner 1983: 95)

At this point culture is no longer a characteristic of people who share a language or ethnicity, it is a broader category used to describe the cognitive features of industrial society. This is an odd shift, not least because it comes just one page after ridiculing the Marxist idea that the proletariat had no fatherland, or that they might share a culture (Gellner 1983: 94). If an industrial culture expressed in different linguistic idioms is plausible, why not a proletarian culture? I am in favour of this broader usage, but we need to be aware of the shift from his dominant usage in terms of language or ethnicity. When Gellner analyses the circumstances under which the transition to industrialism is conflictual, he appears to conflate the cultural differences represented by ethnicity with the broader cultural transformations associated with modernity.

In the chapter on a typology of nationalisms he suggests that conflict occurs most readily in the early stages of industrialisation, and where class and ethnic cleavages coincide. This argument emerges in a number of passages:

The evidence seems to indicate that the classes engendered by early industrialism ... do not take off into permanent and ever-escalating conflict, unless cultural differentiation provides the spark. (Gellner 1983: 93, 95)

It was the social chasms created by early industrialism, and by the unevenness of its diffusion, which made [nationalist conflict] acute ... Whenever cultural differences served to mark off these chasms, then there was trouble indeed. When they did not, nothing much happened ... Classes, however oppressed and exploited, did not overturn the political system when they could not define themselves 'ethnically'. Only when a nation became a class, a visible and unequally distributed category in an otherwise mobile system, did it become politically conscious and activist. Only when a class happened to be (more or less) a nation did it turn from being a class-in-itself into a class-for-itself, or a nation-for-itself. Neither nations nor classes seem to be political catalysts: only nation-classes or class-nations as such. (Gellner 1983: 121)

Three points emerge from all this. First, the anthropological concept of culture, now in free circulation in many other disciplines, remains very important and problematic. Despite a series of commentaries and critiques (such as Kahn 1989) this flawed but indispensable term continues to be freighted with an intellectual baggage which cuts across the best-intentioned analysis. One part of this baggage is the way culture continues to be associated with 'a people', and items of culture are assumed to be specific to 'peoples'. This creates serious problems for the study of ethnic and nationalist movements, where the terms used in analysis ('Serbian culture') risk obscuring precisely one of the themes of that analysis – the political process whereby certain cultural features come to be markers of Serbian identity – and social relations are polarised around selected lines of cultural difference. Clearly, there are many situations where linguistic and cultural differences have hardened into sharp social boundaries; my argument is that they exist alongside other groupings and lines of cleavage, and we need the term 'culture' to refer to an aspect of all these relations, and not only the ethnic ones. If we fail to widen the term, if culture is treated always and only as an ethnic phenomenon, 'we make nationalists a present of their own ontology' (to use Gellner's own phrase, 1997: 94).

Secondly, Gellner's account of the transition reveals that in many cases industrialism and the re-shaping of political and ethnic boundaries are disruptive but not particularly conflictual: people migrate, acquire new skills, develop old and new cultural forms. In his account nationalism generates major conflicts only when ethnic boundaries coincide with class boundaries; in other words we should understand nationalism not as the immemorial struggle of a people, but as a phenomenon which emerges in a recent period in association with the emergence of new social boundaries and rapid economic change. These changes produced new forms of productive capital, new forms of wealth and knowledge; they destroyed and devalued other forms and led to dislocation across the social spectrum. Some have analysed these processes in terms of capitalism and uneven development, some in relation to industrialisation. As grand theories attempting to produce global historical generalisations or identify a prime mover in the rise of nationalism, these approaches diverge. When we look at the more fine-tuned historical or ethnographic analyses of nationalist movements, there is not such an unbridgeable gap between recent interpretations working within Weberian or Marxist traditions. Gellner probably needed the joke about the wrong address at this point in his argument, as a smoke-screen to hide considerable convergence. From a different perspective, Hroch (1998: 106) remarks,

I consider [Gellner's critical objections to my book] a demonstration of his efforts to distance his explanation from Marxism, to which in his historical materialism he was methodologically (though not politically) closer than most of the authors who have dealt with the problem of 'nationalism'.

Thirdly, on the issue of class politics, there *is* an unbridgeable gap between the Weberian and Marxist interpretations, and in my view Gellner weakens his account of nationalism by trying to develop a general theory of conflict in modern society, and by treating class movements as an epiphenomenon of nationalism. In dealing with class conflicts he seems to be operating with only two kinds of political outcome: either there was a successful revolution, or 'nothing much happened'. Revolutionary failure is explained by the absence of the kind of political zeal which only ethnic differences can generate. This is not just an eccentric view of the circumstances leading to revolutionary success or failure – it leaves much twentieth-century political history totally in the dark.

THE ANTHROPOLOGY OF POLITICAL MOVEMENTS

If we look at successful analyses of political movements, those which illuminate how and why they arose, which provide insights for comparative work, we find that they contain at least three strands. The first is that they contextualise the movements in relation to the major historical processes of the society in which they emerge. Indeed many important features of these movements are normally explained by such social transformations: capitalism and class movements, modernity and nationalism, post-modernity and new social movements. In practice the correlations are not so simple and the task of contextualising is very open-ended. The range of transformations which are relevant to an analysis can be illustrated from material in this book. In urban areas such as Milan and Bilbao we find rapid industrial growth which draws in labour and undermines existing patterns of production. In rural areas, such as Andalusia, southern Italy and later the Balkans, we find the liberal reforms of property rights in land. Everywhere we find the widening of markets, not as a one-off event but as a process which continues into the present, affecting regional economies and particular categories of the population. The consolidation and then the collapse of communist command economies transformed rural–urban relations, ethnicity and everything else in eastern Europe. State centralisation and the development of citizenship have a well-documented impact on nationalism, as have the growth of literacy and compulsory schooling on societies with linguistic pluralism. International processes, such as the Cold War and the growth of the European Union, all affect the political movements we shall be examining. Contextualising will always be unfinished and involve analytical difficulties, but better unfinished than unstarted. There are accounts of nationalism, for example, which are inadequate because they recount it as the millennial struggle of a people in a world unmarked by economic, political or cultural transformations.

The second and third strands of analysis I shall refer to as 'movement and discourse'. These are not new terms or new issues in political analysis, but I am particularly indebted to Sewell's (1990) formulation of the problem in his discussion of the 'making' of the working class. He insists both on the close linkage of these two levels, and on the need to think about them separately, and I shall start with 'movement', since it is simpler.

Political mobilisation requires organisation, and this takes many forms: cells, parties, unions, co-operatives, mutual aid associations, clubs, cultural circles. These are embedded in local patterns of work and recreation, but they are also very often federated into a higher-level organisation which constitutes the movement. This involves co-ordination between local groupings and a system of decision-making – and both require communication. This may be personal and clandestine as in the 'underground' movements; more commonly communication is written, while the movement encourages literacy, needs printing presses, and generates its own communiqués, newspapers, texts and libraries. Activists in any movement spend the greater part of their time in meetings and rallies, or printing, distributing and reading information.

These are the activities and organisations which build a social base; they have their own social composition and political culture; democratic or autocratic ways of operating. These organisations are the base from which strategic action to transform society can be launched.

Under 'movement' we can include the repertoire of political action: demonstrations, tax strikes, workplace strikes, sabotage, occupations of land, factories or city centres, general strikes, burning churches or tax records, guerrilla warfare, armed insurrection. This repertoire is itself an important part of the picture. Historians (Tilly 1978) have shown how it has grown and evolved along with changes in the organisation of production and in the technology available to the state. So the repertoire reflects changes in the field of power, though not without some 'inertia', and obviously varies according to the strategic objectives of the movement. But there is more to it than that. Political movements do not have rigid boundaries, since what they do is create processes of inclusion and exclusion which affect everybody who lives in a particular environment. This involves patterns of socialisation at work and after work, in public and private arenas, linguistic codes, speaking with authority or being marginalised. In the local context the movement reflects and modifies existing social differences, even those which have little connection with the overt purpose of the movement itself. In many of the cases we shall be examining, an atmosphere of generalised contestation leads to radical polarisation. Social divides are created through the movement's action, first through everyday forms of inclusion and exclusion, more dramatically in strategies like the general strike. But the most enduring social divisions are created through violence, as we shall see in a number of the case studies.

The subject of a political movement is a collectivity, actual and potential: its discourse articulates who that collectivity are, and their historical purpose. In the case of class, Sewell (1990: 70) has argued that the emergence of class discourse 'took place by a relatively sudden conceptual breakthrough during a period of intense political struggle'. It universalised traditional solidarities, established solidarity between workers of all trades, and empowered them to make collective claims about property and power (Sewell 1990: 71). The breakthrough emerged through a transformation of pre-existing political and religious traditions; once in existence the discourse could survive the collapse of a political organisation, and of course be spread to other societies. Something very similar can be said of nationalist discourse, which builds on, for example, existing conceptions of state sovereignty and loyalty, or local notions of commonality through blood and kinship, but transforms them. A nation is a people, a collectivity with shared attributes, which claims political rights on that basis. The collective interest alone forms the basis for legitimate government, and the government, in turn, claims that the interests of the nation take precedence over all others in the lives of its citizens. Benedict Anderson has done much to illuminate the overall conceptual shift such an innovation represents – and with others has shown how pervasively this model was borrowed and naturalised once invented.

One central feature of all these discourses is that they are concerned with identity: they define who 'we' are. Identity is a notoriously slippery concept, and 'identity politics' is particularly problematic since it connotes the kinds of movement (based on gender, sexual orientation or ethnicity) which are said to emerge when class politics collapses. It is too large a task to disentangle why some kinds of politics are thought to involve identities and some are not, or the many ways identities have been conceptualised and analysed. For our purposes the most useful approach which has emerged treats identity as a narrative, defining boundaries and oppositions. Narratives position collectivities in social processes, and in time and space. In addition to telling the story of who we are, they contain a 'meta' level – an account of why the collectivity exists, derived from assumptions about why the world is as it is. These assumptions may be explicit and theorised in a Marxist-derived account of the proletariat, more implicit in a discourse which understands nations to be enduring and natural units, and history to be an unending saga of competition and alliance between them. Each discourse thus also has its own historicity, its own time-frames and tacit criteria for what constitute significant events. A communist history of capitalism; an anarchist history of liberty and oppression; a nationalist history of Serbia: each selects and makes meaning in different ways.

Identity narratives can be thought of as organised along two axes. One is biographical: it says who are the Basques, or the working class, through the medium of time, an axis with a horizontal line running through past, present and future. As we shall see, class and nationalist movements give different interpretations to this time dimension. The other axis is vertical; it establishes who 'we' are through opposition and the creation of an 'other' – an area which has been explored extensively in research on nationalism and racism. An identity discourse fuses the two axes – the diachronic and the synchronic – so that it is a history of enduring 'opposition'. However, in certain contexts one dimension or another of identity may become more salient, as when political action focuses on an external enemy, or the movement celebrates its unbroken history and the work of the ancestors.

I first thought about these two axes while listening to a seminar paper based on life-histories collected from elderly people who had come to Britain as migrants. Informants had produced accounts of their natal villages, migration, family reunion, the difficulty of bringing up children, the tension between the desire to return home in old age and the needs of the next generation, together with a concern about the death rituals which re-affirm cultural identity. These narratives were interspersed with anecdotes which spoke of another context – for example factory employment and work culture, the (riotous) getting and spending of money with their British workmates. These were parentheses, sideways glances, revealing the existence of other social identities in a biographic narrative which was structured overall by concern with cultural continuities. I mention this source not because it provides an exact parallel, but as a reminder that identity is multifaceted and contextual, and that the analytical issues raised in this book overlap with other fields of anthropology. It is also a reminder that there are many kinds of narrative, from the lone voice to the official histories compiled by the movements themselves. We shall be concerned with collective narratives generated within a movement, more or less stable over time and more or less standardised in their form – if the movement 'captures' the state, then they become more stable. However, I am aware that the term 'collective' begs many questions: How does a particular narrative become authoritative within a movement? To what extent do people recognise their own experience in that authoritative version or dissent from it? These issues will surface often in the case studies, because they are relevant to the analysis of the movements themselves, but some of the more fine-tuned ethnographic questions will be squeezed out by the space available and the nature of the material we are dealing with.

NATIONALISM AGAIN

Having set out a general framework, we can return to specific questions about nationalist and class politics. Gellner and Anderson belong, in different ways, to a school of analysis which stresses that nations are constructed through a political process, and that this happens in the modern period. It happens through the mobilisation of populations seeking recognition of their political rights (usually, but not necessarily, to become a new state), or through the cultural policies of established states. The rights are those of a 'people', conceived either as a bounded group sharing a common culture, or as a race, or as a mixture of the two. A people are defined by their common identity, and nationhood can only be established historically, through demonstrating what does not change over time. These narratives of the nation often have a foundational moment (a birth through migration, conversion to Christianity, a battle of liberation or subjugation) and then move on to recount a series of events – further victories or defeats against an enemy other for example – which confirm the destiny or transcendent qualities of the people. The apparent paradox of using history to talk about the things that do not change (and there is no other way of establishing what does not change) means that nations have to argue their own antiquity, an issue which lies at the heart of the debates in the study of nationalism. Those who oppose a purely modern origin of nations – a group sometimes called, slightly misleadingly, the 'primordialists' – stress the continuities between present-day nations and much earlier social formations. At this point it is important to make clear what is at stake in this debate, and it is not just any kind of historical continuity. Any group can point to some aspect of its culture – be it language, religion or embroidered waistcoats – which distinguishes it from another group and predates modernity by many centuries. Everybody alive has ancestors going back to the origins of the human species. What is at stake is the historical continuity of a people (either as a society with a unified culture, or as a race), and the historical stability of a boundary between peoples.

Most historians have argued that, in societies which were over-whelmingly rural and feudal, the claim that lords and ruling elites shared a common culture with the peasants was both untrue and subversive of the existing order. In a few political circumstances some version of that claim would emerge, but it became the dominant framework only when Europe was re-organised into a system of states, each of which claimed to be the legitimate manifestation of the rights of a people. In Europe, this concept of a people, and 'their deep horizontal comradeship' – to use Anderson's (1983) phrase – erupted onto the political scene with the French Revolution. Rather than seeing pre-existing cultural forms and divides as providing the blue-print for the map of nation-states, we should see culture as a contested terrain, providing one framework for the artic-ulation of various political claims. In the nation-building process, what we refer to as culture undergoes a subtle shift, from a lived-in series of shared understandings to a collection of politicised representations which embody or symbolise the nation. It is a shift from culture-in-itself to culture-for-itself, as Gellner might have said.

The politicisation of culture occurs in the establishment of both the boundaries of the nation (the 'vertical' axis of an identity narrative) and what constitutes the common culture of a people (which tends to be established through a horizontal time-line of historical continuity). In fixing external boundaries there is a great deal of flexibility and opportunism. What cultural features made the people of the southern

Tyrol Italian but the people of Savoy French – much to Garibaldi's annoyance, since he was born there? Internally, the task of delimiting a common culture led some forms of nationalism, particularly in eastern Europe, to re-work and recontextualise peasant or folk culture as the core of their identity. Others, like Italy, were originally established 'against the peasant' and elaborated a national identity around an elite culture of classical civilisation. As Beissinger remarks (1998: 175), 'Nationalism is not simply about imagined communities; it is much more fundamentally a struggle for control over defining communities – and particularly a struggle for control over the imagination about community.' What counts as national culture is not some totality, but the parts which are distinctive; not static, but the result of competition between various groups to define the key experiences.

As I indicated at the beginning, the research agenda on nationalism is very rich, and my intention in the case studies is to emphasise certain strands within it. The main problem is that analyses which stress that nations are constructed have to work against the grain of phrases ('French culture', for example) which imply that 'cultures' can be mapped quite simply onto 'peoples'. Using the term 'constructed', or saying that they are 'imagined', should not be thought to imply any lack of reality in either state boundaries or the processes through which national identities emerge. It is simply a way of stressing that accounts of the emergence and evolution of national identities have to be rooted in a social and political context: how linguistic or religious differences within a society are embedded in social relations, how they co-exist with other sets of differences, how they can become ethnic markers. In order to understand the forces which set nationalism in motion we need to look at a broad range of issues – shifts in the economy, state centralisation, cultural policy – and their effects on specific groups: the response of local political elites, the fate of peasants in the Basque country, or of the young urban unemployed as the Yugoslav economy collapsed in the 1980s. We know that nationalism appears in periods of change, and that it combines old themes in new projects, but beyond that generalisation is difficult and sometimes merely reductive, since what appears crucial is the interaction between these long-term processes in specific social settings. That said, in the conclusion I shall explore some specific arguments about nationalism and the demise of rural society within Europe.

CLASS

In the social sciences and outside them, class denotes a very wide range of phenomena. The most neutral definitions refer to 'structured economic inequality' (Coole 1996: 17), acknowledging that this covers a variety of theoretical positions. Within the Marxist tradition class generally refers to a structural position within a production system; other sociological traditions employ a broader usage of class as a category of people defined by their occupation or income levels, while at the furthest extreme from the Marxist tradition we encounter a purely statistical model of social stratification. These usages are themselves embedded in different analytical concerns, from mapping patterns of consumer behaviour to explaining the reproduction of structured inequality within society or the dynamics of capital.

'Class politics' has almost as wide a range of referents, from the micro-politics of snobbery to revolutionary drama. As a term it can include the way governments increase or decrease structural inequality through taxation and education policies, or the way in which people create, accommodate to, or resist such inequality in their daily lives. Bourdieu (1984) for example has been concerned with class politics in the sense of a diffuse system of discrimination and exclusion centred on lifestyles and cultural capitals. This is significantly different from class politics as the large-scale mobilisation of people to transform the organisation of society and its government. All these overlapping connotations and concerns are bound to create confusion, so it is worth clarifying at the beginning that this book is primarily concerned with these class movements and only tangentially with other processes.

Unfortunately this clarification does not tidy away all the definitional problems, since analysis of class mobilisation inevitably evokes 'a class' (as yet undefined) which is mobilised. It involves questions about the relationship between economic and political processes, and the shifting definitions of class in our account of this relationship. This emerges most famously in the Marxist formulae about how a 'class-in-itself' can become a 'class-for-itself'. A structural position, defined analytically as part of an economic relationship, becomes a category of people (the proletariat) who then acquire consciousness and become political subjects. In fact the debates about class-in-itself and class-for-itself have raised some of the most enduring issues found in the social sciences: determination, structure and agency, the existence of historical laws. Within that debate some have argued that classes are 'given' as an economic category by the organisation of production, and that the composition of the political movement is congruent with the economic category, give or take a few intellectuals who are necessary for generating consciousness. And that is the point: class as a political term is an economic category with added consciousness. Others want to break with these more reductive models, arguing that class mobilisation is a response to a variety of processes – to not only the divisions created by systems of production, but also political and cultural processes which are not derived from some universal logic of capitalism. Class movements interpret economic processes, and as a result they create solidarities in different ways, so that their composition varies and their political agendas go beyond what is conventionally referred to as the 'economic'.

In this second body of writing, we find the argument that rather than classes pre-existing their mobilisation, class is constructed through mobilisation, in a society marked by structured economic inequality. Class, in this sense, is constructed politically, and hence, in part, is constructed 'discursively'. Marxism is doubly important in this context, because it gave rise not just to theoretically sophisticated analyses of capitalist society, but also to very powerful political movements. Any attempt to analyse class movements in the twentieth century therefore has to deal with the impact of Marxism on those movements, and a small step in that direction is taken in two of the case studies. It is a difficult task, not least because of the long shadow Marxism has cast amongst social scientists. Many, having moved away from the theory because of its teleological elements, have also become completely uninterested in the kinds of phenomena which they were once trying to explain; others, who were never themselves Marxists, have perpetuated a very narrow view of the scope of these movements, and a reductive account of the relationship between economic and political organisation. One place where this becomes a problem is in disentangling arguments that the life and death of class movements are linked directly to the fate of the industrial proletariat. In fact, historical studies have shown that class movements were rural as much as urban, inspired by anarchist ideals as well as by socialism, and mobilised not just wage-labourers but a broad spectrum of economic actors, in a variety of emancipatory programmes.

By keeping sight of this variety we will also make better sense of 'new social movements', an over-generalised category which first emerged in arguments about the death of class and what was to come next. There has been a romantic tendency to portray all these movements as progressive, whereas many of the most important recent political developments in Europe have involved direct action to defend corporate interests, the mobilisation of racism, and opposition to immigration. Some do represent a break with class politics, attempting to create collectivities and coalitions around a new political agenda; others are a continuation of political strategies and aspirations which were common in class movements in Europe and elsewhere. The older revolutionary movements are unlikely to return to the political arena, but revisiting them can still give us some critical understanding of the world in which we live, and form a base-line for assessing the new strategies and coalitions at work in contemporary anti-corporate and 'anti-global' movements.

These comments have implications for the three strands in the anthropology of political movements outlined above. The analysis of class movements has always involved the first strand – that of contextualising them in relation to long-term economic processes – since this has been the dominant mode of explanation, sometimes deployed in a rather mechanical way. I shall use a broad range of historical and ethnographic studies, chosen to demonstrate that class politics was not synonymous with the history of wage-labour, or of manufacturing industry. They also allow us to explore the different levels at which these movements operate, and the way they shaped each other. For example, 'the working class' as a political entity was never a simple aggregate of local workers' organisations or communities of resistance, since the national and international levels fed back down to modify the boundaries, strategies and objectives of class action.

By broadening the picture we also open up the second and third strands of analysis, which involve the political construction of class through mobilisation and the variety of discourses which define class identity. These include ideas about how people realise themselves through work, or how their social being is defined through work relations. The movements were certainly active on wages, piecework and the kinds of issue which occupy present-day trade-unionism, but these were combined with a more radical programme on property rights and interpretations of the relationship between owner and labourer which reversed the normal notions of dependency. They were also concerned with the public sphere and social justice, contesting control of public spaces, developing critiques of church teaching on gender, and of many other forms of social hierarchy. This is one of the reasons for their enduring interest.

There are many ways of dealing with the cultural aspects of class movements. I have concentrated on the identity narratives, and on the extensive critiques of existing social relations in which they are embedded. I have given less space to discussion of their symbolic and ritual activity. It seems to me that symbolism is sometimes given prominence because it appears to offer a way of dealing with the emotional aspects of political life, in a framework where (to return to the opening comments) interests and passions are seen as alternative and opposed motives for action. It is not necessary to portray these passions as an irrational dimension in political life, requiring separate treatment – as though emotions start when reason stops. Red flags are defended and fought over: they are profoundly moving to a movement's supporters; they make the movement visible as songs make it audible; they speak of harmony and presence, power and defiance. They are the focus for solidarity, and sometimes for divisiveness, but solidarity is not an end in itself, and when solidarity collapses it is not a failure of symbolism. Symbols are indeed important, but we shall grasp them better if we put them back into the rich 'counter-culture' articulated by each movement and into the context of purposeful action.

In sketching a common framework for the study of political movements, there is no intention of flattening out the important differences between them, even if these differences do not always lie in the factors suggested by the canonical texts. At the outset we can suggest a number of themes which provide a *provisional* set of contrasts. The nationalist movements attempt to unify a group of people they consider to be related through descent, in order to secede from the existing state. Their discourses stress homes and homelands, and their identity narratives are strongly historical, established through continuity between the present and the past. Class movements create horizontal solidarity in order to achieve a transformation of the existing state, or to abolish it. Their discourses stress relations established at work and in the workplace, and their identity narratives have a strong future orientation: the workers are defined around who they will become. These contrasts will need some qualification as the ethnography unfolds and new themes emerge. The similarities derive from the fact that in each case we encounter people living through periods of rapid change and dislocation who forge an identity and an interpretation of history which makes their own values and experiences central in a narrative of how society should be, and forge a political strategy to make that happen.

As stated earlier, the argument of this book is built up through case studies which allow us to explore these political movements as the inter-section of a variety of social and cultural processes. My intention is that each chapter should be free-standing for certain purposes – useful for those seeking information on a particular area – while the comparative and cumulative issues are brought together in generalising chapters and sections which intercut the case studies. Other ways of organising the material, such as surveys or thematic summaries, would have allowed the inclusion of more examples, but put a different kind of strain on the reader – and in fact pulled apart the interconnections which are central to the overall analysis. No one volume could deal with class and nationalist movements comprehensively on a pan-European basis; that said, a case-study approach is selective in a more obvious way.

The material analysed comes from various periods, from the beginning of the twentieth century to the present day, and from contiguous societies which run in an arc round southern Europe from Spain through France and Italy to Yugoslavia. This choice has been shaped positively by the strengths of the available literature, and negatively by the limits of my own knowledge. It is only one possible 'cross-section' of Europe, but it is a particularly interesting one. It contains within it examples of virtually all the major political currents of recent European history, from fascism to anarchism and communism, with Christian Democrats, liberals, populists and many others in between. These are varied and dramatic political environments, dominated by movements which made sustained attempts to transform the state, and which questioned, from a variety of perspectives, the legitimacy of existing governments to represent the people. Some themes crop up repeatedly in these studies, most notably the political responses to the crisis of rural Europe in the twentieth century. By taking a more limited range of cases (for example, by excluding Yugoslavia and the excursion into France) it would have been possible to engage with the work of those who have argued for a particular south European pattern of economic and political development (Hadjimichalis 1986; Hudson and Lewis 1985). The strategy here is different: to explore a varied cross-section in the hope that it has wider analytical relevance, rather than demarcate a region and build an argument around its specific history. The reasons for this choice will be clearer if we look briefly at the relationship between theoretical paradigms and the emergence of regional specialisation within anthropology.

ANTHROPOLOGY AND REGIONS

The central fact here is that what anthropologists find, in this or that place, far from being independent data for the construction and verification of theory, is in fact a very complicated compound of local realities and the contingencies of metropolitan theory. (Appadurai 1986: 360, quoted in Fardon 1990: 24)

We notice, however, that for Kapferer, the only 'conditions' that are seen to be operating upon the formation of knowledges are those of being 'inside' (as the native is) or 'outside' (as the anthropologist is) the culture under investigation. (D. Scott 1994: 129)

The great nineteenth-century social theorists generated a series of conceptual dichotomies in their attempt to understand the ongoing trans-formations of Europe. By the twentieth century these had, in turn, fed into a division of labour within the social sciences which left anthropology responsible for the task of understanding 'pre-capitalist', '*gemeinschaft*', 'traditional' social forms, which it did primarily through methods employing 'one fieldworker plus tent' (Hann, 2001). While this period lasted, British and American anthropology did not occupy itself with Europe. Only in the 1960s, through a series of conferences and publications, were the scattered studies of rural life consolidated and rendered visible through the construction of the Mediterranean as a region. Anthropologists still stayed on one side of the traditional–modern hyphen, concentrating on the life of rural communities and framing analysis around the search for social and cultural forms (honour, patronage, godparenthood) which were common to the Mediterranean area. In 1977 John Davis published *People of the Mediterranean*, an encyclopaedic overview of this anthropological work (all other social science contributions were rigorously excluded), but its appearance more or less coincided with a complete shift in the scenery.

After that date anthropologists continued to study moral systems and localised practices involving gender, status and the person, but in ways which became more sophisticated when freed from the assumption that these were in any simple sense 'traditional' or had to be analysed as part of a Mediterranean culture. Alongside them emerged research which in effect reconceptualised the 'micro' and the 'macro', studying local processes in relation to capitalist development, state formation or Vatican reforms. There were strong critiques of assumptions that bounded communities constituted an adequate framework for analysis, or that fieldworkers were studying lands that time forgot. In a variety of ways anthropologists engaged with historical materials in the interpretation of contemporary realities. Much of the renewed ambition and anthropological innovation of the 1970s derived from the work of Eric Wolf and his students: out of this and other contributions came new frameworks, such as centre–periphery relations and Europe itself as an analytical category. There was nothing monolithic about the results: both theory and research moved in a number of directions. European social science traditions started to make an appearance in the English-speaking world, while some established scholars broke new ground, and some younger scholars attempted to revive the Mediterranean paradigm and the study of communities. As a result the developments of the last 25 years can be told from a variety of perspectives: the introduction to Goddard, Llobera and Shore (1994) is a very useful overview, and gives references to many of the other relevant historical summaries. Here I want to make two points which are specific to the issues raised by this book.

First, the emergence of a region as a focus for anthropological inquiry builds on empirical and theoretical assumptions about the priority of certain social and cultural phenomena. 'Regions' make some things central, defining; others marginal or invisible. From work within the Mediterranean paradigm we learned a great deal about certain kinds of cultural phenomena; we learned very little about colonialism, urban industrial centres, or the politics of religious or class identities. The later emergence of an anthropology of Europe marks not just a geographical redefinition, but a shift in analytical focus, not least towards processes of economic and political integration. However, the issues are defined primarily around processes within western Europe and the European Union; eastern Europe remains another kind of specialisation, at least until the collapse of communism. We should also note that, 20 years after the first moves to establish an anthropology of Europe, we still know more about the subtle sociological variations between Greek villages than we do about Germany. Regional generalisation involves selectivity in research and the emergence of a consensus about the centrality of certain phenomena in the analysis of why a society is as it is. These paradigms go with the territory; that does not make them valueless, provided that we are aware of the limits of their explanatory power, and that for certain kinds of research we shall have to break with them.

This leads on to the second point: for anybody interested in political movements there is a particular problem with this anthropological literature because of the way it deals with culture and identities. There has been a very strong tendency to concentrate the analysis on aspects of culture which belong to territorially-based entities (like the Mediterranean), or 'peoples' in the sense of nations and ethnic groups. The focus on social entities, which are indeed defined by their common culture, is one of the strengths of the anthropological tradition, but it can also lead to a narrowing of the range of cultural phenomena analysed, in a way which vitiates even the study of nationalism. If culture is invariably connected to 'peoples' then, as noted above, we risk naturalising nationalism, and are likely to marginalise the study of other political movements built around other identities and other kinds of culture. Some of these have considerable historical importance, such as the development of Catholic social doctrine and Christian Democracy, or the transnational peasant populism of eastern Europe.

It is in the study of class politics that the anthropological paradigms around culture and identity become most problematic. One example we shall be dealing with is the anarchist movement in rural Andalusia, a form of politics which is so unlike much of the European 'mainstream' left that many writers have dwelt on its irrational or millennial characteristics, or the archaic social conditions which must have generated it. But anarchism was not just found in Andalusia, and a short summary of a movement in southern Italy will allow me to make one concluding point about the purposes of this book.

After Italian unification, property in the central plains of Puglia became concentrated in latifundia, while three-quarters of the population were landless labourers resident in sprawling urban centres (the 'agro-towns'). The labourers worked 'from sun to sun' for the intense months of the wheat cycle and then suffered unemployment. Anyone who wants evidence of the depths of routine misery and brutality to be found in parts of southern Europe in the early twentieth century should read the description of a 'Company Town' in Frank Snowden's monograph *Violence and Great Estates in the South of Italy*. But as in Andalusia it was wealth, and its conspicuous concentration, which spurred political radicalisation.

Snowden (1986: 116) says that the resulting political movement was socialist in name, while syndicalism remained the substance. The workers supported the Socialist Party at elections, but remained a radical current within it: they were highly critical of the reformist tendencies of the parliamentary leadership, who in turn considered the southerners backward and impulsive. More important than the party were local sections (*leghe*) which organised employees, and the chambers of labour, which built bridges between the struggles of the (temporarily) employed, and those of unemployed men and women. As in Andalusia, this dual structure mobilised opposition in the workplace and on a territorial basis. The general strike was adopted as the most effective form of action, halting all economic activity and sealing off the town even for the transport of food and water. Even the mobilising slogan, 'All of us or no one' echoes those of Andalusia.

Strikes in the early part of the century found the landlords unprepared – they either capitulated or responded with violence, using private strong-arm men and the army. Both outcomes further radicalised the movement (Snowden 1986: 100). There were significant gains on wages and working conditions, but the revolutionary goal remained the social-isation of the land. Alongside this, as Snowden (1986: 109) insists, were aspirations for moral reform and social equality; indeed, in the short run, local activists gave them priority. This involved social advancement through literacy classes, curbs on alcoholism and delinquency, and the pressing of the claim to equality as human beings through the occupation of public spaces and a break with deferential forms of behaviour.

Feminism was an integral part of the politics of anarcho-syndicalism. Hierarchy in any form was anathema to its libertarian ethos. A continuous effort was made to raise the consciousness of women's rights, to encourage women to take an active political role ... One of the most effective speakers the movement possessed was the anarchist feminist Maria Rygier, whose oratory was so powerful that even landlords came out of curiosity to see her. (Snowden 1986: 114–15)

By 1920 the movement had strengthened and was co-ordinated on a regional basis, and represented a direct challenge to the power of the state. As in other regions, the landlords put their resources into the fascist movement. Repression took longer than in the north, where the more centralised left-wing movements were decapitated quickly, but a combination of contract killings, fire-bombs and *squadristi* working in open collaboration with the police insured that, by the middle of 1922, here too no opposition remained above ground (Snowden 1986: 175–202).

Two countries, one political culture. However one handles the reasons, there is a remarkable convergence between Andalusia and Puglia, in their rural hierarchies, their forms of political action and mobilisation, and their moral understandings and representations. In the beginning anthropology went looking for elements of a pan-Mediterranean culture common to peasants, shepherds and townspeople, and located it in honour codes, godparenthood, and aspects of spiritual belief. The kind of convergence annotated above escaped the net. As the Mediterranean research agenda lost momentum, national and ethnic particularity took over. Reading certain monographs, it is as though the anthropologist's informants woke up every morning asking 'Who am I?', to which the only conceivable answer was 'I am Greek' (or Romanian, or Irish). When attention shifts to large-scale political movements – and it is rare unless they are based on ethnicity – the analysis is so imbricated in local usage, linguistic particularity and symbolic subtleties that it is hard to see how such a movement could ever have emerged except in that specific context. Incomparability reigns supreme and the anthropologist translates between cultures. There are some important exceptions – for example Aya (1975), who compared political movements in Sicily and Andalusia, and the very important contemporary work by Holmes (2000), whose analysis of pan-European 'populism' will be discussed in chapter 8. One of the purposes of this book is to build on this and other research in order to loosen the anthropological concept of culture from its assimilation to ethnicity, and re-explore alternative bases for comparison.

ORGANISATION AND THEMES OF THE BOOK

The themes of the three chapters on class movements have been chosen to show their diversity. Chapter 2 deals with the evolution of political action amongst industrial workers in northern Italy, using sources on Sesto San Giovanni near Milan and on Gramsci's Turin factory councils prior to fascism. It explores the forging of class identity in a labour force composed of artisans and 'unskilled' labourers, and the impact of

Marxism on that process. It is here that we first encounter the importance of the future tense and the revolutionary road in the narratives constructing class divisions.

Chapter 3 looks at the anarchist movement in rural Andalusia from the mid-nineteenth century to the Civil War, a period characterised by extraordinary economic stasis and the repeated use of military force by the Spanish state. It begins with the moral absolutism of the struggle and the discursive connections which were made between revolution and destruction. It then goes on to analyse the organisation of production on the big estates, their relationship with the *pueblo*, the social bases for solidarity and general features of anarchist political practice.

Chapter 4 continues the analysis of identity and the revolutionary road in relation to a period of class mobilisation which did not involve a proletariat – that of the share-cropping movement in the red belt of central Italy that was led by the Communist Party. It explores the strategy of a western Communist Party during the Cold War, illustrating the way global politics are incorporated into and transform local struggles, and the long-term consequences for the share-croppers themselves. It concludes with a commentary on the attempt to achieve a revolutionary transformation through strategic reforms in a democracy, and the increasing stress on identification with the political traditions of the party in the period from 1945 to 1990.

Chapter 5 returns to the theoretical issues arising from using class as an economic and political category, and then reviews the convergences and divergences which emerge in the three examples. It is organised around the themes of movement and discourse, and discusses in particular the contrasts between communist and anarchist political cultures.

The next two chapters are extended analyses of two ethno-nationalist movements – those of the Basque provinces and the former Yugoslavia. There is more research on this kind of politics within anthropology, and the examples were chosen because of their contemporary political importance, as well as to provide coverage of both eastern and western Europe – a divide which emerges frequently in the theoretical literature. Chapter 6 analyses Basque nationalism from 1890 to 1990 on the basis of a very rich and voluminous literature. The first part traces the emergence of a narrative of loss amongst the urban intellectuals and the petty-bourgeoisie, and its incorporation into the social and moral map of the rural population. This leads to a discussion of ethnogenesis (the formation of a group defined ethnically), and some of the problems with the anthropological category of ethnicity. The final part deals with the post-Franco period and the problems ETA has posed for a democratic government by its strategy of violence and armed insurgency.

Chapter 7 analyses the wars in Yugoslavia in the 1990s, in relation to the complex patterns of economic and cultural differentiation within Yugoslav society in the twentieth century. This is a shorter but broader background than the standard '600 years of Serbian history', and involves setting out a number of themes. We shall look at how cultural diversity emerged in the practices involving kinship and neighbourhood, how the state recognised and institutionalised nations and minorities, and how both of these levels of cultural difference were affected by the transformation of peasant/urban relations, as well as by the development and collapse of a 'command' economy. The chapter then discusses the political strategies of the dominant groups which emerged in the late 1980s, and the reactions of the 'international community' to the collapse of communism. The most general issue raised by the chapter is once again the problem of ethnicity, but it emerges out of an examination of the way in which the term has been employed by political commentators and the media as an explanation of violence in the Balkans.

Chapter 8 returns to general questions about nationalism, and pulls together the case material on nation-building – as a political process, as social transformation, and as a cultural framework. It then looks at whether the themes which have emerged as characteristic of class and nationalist movements can to some extent be combined. Evidence for this comes from the existence of movements which articulate a potent mixture of class and regional or ethnic identity. The rest of the chapter looks more briefly at two movements (in southern France and northern Italy) which have this hybrid quality, and raises questions about populism and the dynamics of contemporary European politics.

Chapter 9 is a conclusion which draws some themes together, both around the anthropology of political movements and the politics of southern Europe. We shall go back to the question of identity and narratives, and to the way, for example, that an over-arching discourse of class or nation 'speaks to' the experience of people in local settings. There are also interesting parallels in these movements which derive from the politics of polarisation. However, I think that the most interesting result of bringing this material together is the possibility it offers to ask questions about the process of modernisation, and the concept of modernity. One substantive dimension of this is the demise of rural society – a much more recent phenomenon than most people (especially in Britain) realise, documented in Table 9.1 on p. 188. This transformation has had a profound political impact, and Nairn (1997) has explored this in a recent article on nationalism. Though critical of his representation of rurality, I think we are all indebted to him for the brilliant introduction of the symbol of Janus into our thinking about nationalism and modernity. There are 'backward-' and 'forward-looking' aspects to all these movements, in the struggle over autonomy in production as well as in the more familiar evocations of rural idylls. There are also intricate questions about the way the past is evoked in these discourses, about what changes are represented as progress, and whether they are reversible.

A final note to the reader: in writing this book, one of the trickier decisions concerned a style of presentation which would give adequate references while making the text as streamlined and accessible as possible. I have dispensed with footnotes and avoided over-meticulous references to every theoretical discussion in a field which already contains a number of synthesising studies. Instead there is a short guide to sources at the end of each case study and a cumulative bibliography at the end of the book.

2 NORTHERN ITALY: 'A WORLD TO WIN'

Sesto San Giovanni was a suburban village with a number of artisan workshops and small silk mills, five miles north of the centre of Milan in a fertile farming area. In 1840 the railway linking Milan to Monza was built, and in 1882 the tunnel under the St Gotthard Pass provided a rail link to the territories north of the Alps. Power came down from the mountains – hydro-electric power which could now be transported long distances. By the end of the century, the population of Sesto had grown to 7,000 inhabitants, still mostly employed in farming or in small-scale workshops, but it had the space, location, communications links and supplies for a major industrial centre. Between 1900 and 1914 many of Italy's most important industrial firms moved into the district – firms like Falck, Breda, Marelli, Campari, Pirelli – with a particular concentration in the engineering sector. On to the old nucleus of Sesto was grafted a new town of giant factories and workers' houses, built in a grid pattern and often owned by the industrial firms. By 1920 the resident population had tripled to nearly 20,000 inhabitants, but this was only the beginning. By 1944 the engineering firm Breda alone employed 12,000 workers. Growth continued in the years of Italy's economic miracle in the 1950s and 1960s, and although many firms then collapsed or relocated, at the end of the twentieth century Sesto had a population of nearly 100,000.

If Sesto has a hundred-year history as one of the primary industrial centres of Italy, it also has a long tradition of political radicalism which began immediately after the First World War. This was a period when the demobilisation of conscript armies and the shock waves from the Bolshevik revolution generated political uprisings across Europe, from Andalusia to Bavaria. In Sesto the Socialists captured the town council in 1920, at the beginning of two years of tumultuous political activity (the *biennio rosso*), of strikes, demonstrations and factory occupations. The fascist reaction began with the brutal killing of left-wing activists in Sesto. Under fascism there continued to exist a militant communist underground, while the partisan organisations during the later stages of the Second World War earned Sesto the epithet of 'Italy's Stalingrad'. The Communist Party emerged as the dominant political force in the decades after 1945.

I have chosen to start the case studies with Sesto because it is precisely the kind of place where we would expect class politics to emerge: a major industrial centre with a concentration of factory workers, a place which developed enduring and evolving forms of class action. Yet the fact that this process is at one level both familiar and apparently predictable may lead us to see class movements as a direct consequence of certain kinds of economic organisation. This chapter will try to challenge this sense of inevitability for two reasons. First, we need to understand the political work that goes into constructing a class movement, work which has to overcome many kinds of internal division and can lead to a variety of political forms. Secondly, the features which lead us to take for granted class mobilisation in places like Sesto – towns, factories, wage-labourers – are missing in the other movements we shall be looking at, so clearly we need to widen our perspective on what is happening in order to develop comparisons.

This chapter explores these issues through the discussion of the dual themes of movement and identity. The first part is concerned primarily with the organisation of factory workers, who came from very different social backgrounds and were internally divided by their conditions of employment. Factories had the potential to create solidarity or competition amongst employees, and if solidarity eventually triumphed it owed much, as we shall see, to organisations created outside the workplace. The second theme is identity, and specifically what made somebody a member of the working class. In this environment artisans and the new 'unskilled' labourers had very different attitudes and experiences of work relations. The narrative of class identity which came to dominate was woven out of a variety of conceptions, and went beyond simply delimiting those who shared economic interests. It constituted a powerful account of history and morality, heavily oriented towards the future and with the concept of revolution at its heart.

FACTORIES AND LABOURERS

Assembling a labour force was a constant preoccupation of the early entrepreneurs and managers. A town which doubles its workforce every decade or two is a town of migrants, and this implies, in a country as recently and partially unified as Italy, the presence of considerable social and cultural diversity. The farmers and silk workers of the old town were joined by skilled metalworkers recruited from the centre of Milan, from

the Alps and from other centres in Lombardy, as manufacturing was concentrated in the Milanese periphery. Prior to 1914, many came from
outside Lombardy, from Tuscany and the Romagna, while later in the
century the main influx was from southern Italy. They came from rural
and urban environments, speaking a range of mutually unintelligible
dialects (Bell 1986: 20), and from regions with very diverse religious and
political cultures. For all the migrants, whatever their linguistic and
religious backgrounds, the move to factory employment represented a
cultural as well as an economic transformation of their lives.

The workforce was marked by a strong gender segregation and
hierarchy. Willson's study of women workers at the large Magneti
Marelli plant during fascism shows that it was extremely rare for men
and women to be employed in the same workshop, and that there was
very little direct competition between them for jobs. This pattern was
already apparent in the formative years prior to 1920. In a factory like
Marelli, with a great deal of assembly work, women accounted for nearly
half of the workforce (Willson 1993: 103), whereas they were virtually
absent from the heavy engineering plants. They very rarely formed part
of the craft leagues; they were employed in those sectors of production
which did not require great physical strength, long training or education.
As in so many other situations, women were seen as suited to jobs which
were repetitive and boring, but which required considerable stamina and
dexterity. They were paid wages which were only a half or a third of those
paid to men (Willson 1993: 98), even though there was a considerable
number of female-headed households. The information available on
household budgets, and on the effects of grading work by gender in terms
of wage costs and social relations, is very limited for the earlier period.

The most important source of variation amongst the factory labour
force is the distinction between skilled and unskilled workers, a
distinction which itself is partly articulated within a gender discourse.
The categories 'skilled' and 'unskilled' are obviously problematic. The
distinction is found within workers' associations and in management
practice, and as industry expands it is managers who decide what does
and does not constitute a skill, and how it should be remunerated. Despite
the problematic character of the terms, I shall use them as a shorthand
for a distinction which is central in the following analysis, not least in
the difficult questions about the construction of class identity.

There were two main categories of people recruited to the factories of
Sesto. First were skilled men, who had been through an apprenticeship
and craft training: as metalworkers, printers or masons. Historically
these skilled workers had been independent artisans, selling not their
labour but the product of their labour, setting their own work rhythms

and providing their own guarantees as to the quality of their goods. For this category of worker, wage-labour employment was sufficiently recent and uneven to generate areas of friction in the factory. According to Bell, entrepreneurs showed greater interest in training skilled operatives from urban workshops than in recruiting the unskilled of the city and surrounding countryside. By 1911, 58 per cent of Sesto's workers were from skilled backgrounds. There was less direct recruitment of rural workers in Sesto than in other industrialising areas, and although 'peasants' of course had many skills and their own work culture, these were largely irrelevant to the needs of industry.

Whatever their origins, the workers had to be inducted into a new production system, with new machinery and labour processes. Old skills, hierarchies and work teams were broken up, while new mechanisms for supervising and accelerating work rhythms were instituted: 'To build a standardised labor force industrialists adopted a detailed set of factory regulations which aimed at undermining the craft mentality of skilled artisans and at conditioning unskilled personnel to modern manufacture' (Bell 1986: 23). These factory codes regulated hours of work, forms of dress and behaviour, responsibilities and management structures. Management was constructed around a new hierarchy of engineers and foremen, replacing the collective workshop teams characteristic of artisanal production.

This was a period of rapid innovation. Pirelli and Breda, after sending commissions to the US, were pioneers in introducing assembly-line techniques into Italy. Rationalisation meant the breaking-up of old production techniques into component operations and then re-assembling the product; new forms of knowledge and skill were required of the workforce, old ones made redundant. Much research went into organising the labour force to optimise the productive capacity of machinery: the classic time-and-motion studies with a stop-watch to establish the baseline for work rhythms. In *The Clockwork Factory*, Willson suggests that Marelli was the firm most committed to the intro-duction of Taylorism, and one of the striking facts to emerge from the study is the astonishing number of white-collar clerical workers that full-blown Taylorist management required – by 1942, 20 per cent of Marelli employees were clerks. Piecework was the most widely used system of wage-labour remuneration in Sesto factories. It was judged the best way to achieve work discipline and maximise production. It both required and constituted the motivations necessary for the new production system to run competitively.

All these developments outlined for Sesto – the disintegration and reconstitution of labour processes, the fragmentation and control of the labour force – are familiar from other accounts of early industrialisation and complicate any argument suggesting that class solidarity is generated by the simple facts of economic organisation. Although workers increasingly shared similar conditions of employment, there remained substantial obstacles to unity and mobilisation amongst this emerging industrial proletariat. The cultural and political diversity of the workforce was compounded by the distinctions in the factory: between men and women, between clerical, skilled and unskilled workers. They were divided between different companies, each with its own regulations, emerging corporate cultures and policies on housing, welfare and recreation. Within the factories there were often strongly demarcated territorial and social divisions between the various plants, while the noise levels – especially in the engineering sector – made any form of verbal communication impossible. (There were complaints to the mayor that the new Falck steam hammers were shaking the whole neighbourhood – Bell 1986: 233.)

Common conditions of employment may create competition, and be deeply divisive, impacting in different ways on workers from different backgrounds:

Skilled workers especially objected to piecework, perceiving it as a derogation of their tradition. Craft training, however, was of little use in resisting such altered work-loads, and craft-based organizations produced little labor solidarity... Less-skilled workers, by comparison, often welcomed piecework as a means of increasing income, an attitude which led to conflict between various labour categories. 'This system generalizes hatred among workers', wrote *Sesto Lavoratrice* (a local journal) in 1911, when a number of Marelli lathemen complained of antagonisms over piecework between skilled and unskilled workers. 'Piecework', the journal concluded, 'is an extremely widespread practice and a hateful means of exploitation.' (Bell 1986: 31)

The managers' right to manage was unchecked by any established need to consult or negotiate with workers' representatives, or any substantial forms of state regulation. Breaches of factory rules led to fines, threatening forms of disobedience to dismissal. Work rates were increased and lines speeded up, while major re-structuring was normally accompanied (then as now) by mass lay-offs and the re-hiring of labour under new terms and conditions.

Neither solidarity nor competition are conferred by the facts of economic organisation alone; solidarity also has to be constructed as a political project, on the basis of a particular understanding and vision, in the face of very real divisions and opposition. It took longer to construct than the houses and the chimneys; it was the political work of decades. There were only a dozen strikes in the first 15 years of the century in Sesto; they all appear to have been about job control and work routines, and to have involved only part of the workforce in any one factory. Effective unionisation was slow, while political unity was even slower and more fragile. Only in 1920 did a socialist slate of candidates win control of the Sesto municipality, to be undermined shortly afterwards by the schism which led to the foundation of the Italian Communist Party in 1921.

CREATING A CLASS MOVEMENT

Bell's study of Sesto San Giovanni in the early years of the century is a contribution to a debate in labour history about militancy in the early stages of industrialisation. Bell argues that working-class formation owes more to the actions of craft workers in defence of their trade than to the demands of new masses of unskilled workers, not least because the latter were so fragmented by the factory system. In the process he shifts attention away from the shop-floor issues of pay and conditions, towards the analysis of pre-existing political and cultural traditions amongst artisans, and to political action outside the factory. In this respect there is evidence that Turin and Milan, for example, may have followed rather different economic and political trajectories. However, I am more concerned with the co-existence of different kinds of mobilisation than with the debate about their primacy or relative weight in any one environment.

The characteristic form of artisan organisation was a mutual aid society. These developed rapidly in Italy after unification, and by 1894 there were over 6,700 such associations nationally, with a total membership of nearly one million. The *Societa di Mutuo Soccorso di Sesto San Giovanni*, founded in 1880, had several functions:

to assist injured members, to provide pensions, to pay funeral expenses and to provide for the education of members and their children ... it initiated an evening school [which] eventually provided Sesto factories with trained apprentices and in so doing served as a bridge between the local artisan milieu and that of the industrial craft worker. (Bell 1986: 48)

Bell also notes that such societies were not originally founded upon concepts of class solidarity and commitment to political change; they were devoted to self-improvement and mutuality, and were open to middle-class supporters and benefactors.

However, organisations evolve with the environment in which they operate, and pre-war Sesto was affected by both major political events and booming industrialisation. The riots in Milan in 1898, when more than a hundred civilians were killed by the army, had led to a round-up of labour association leaders and a temporary ban on the mutual aid societies. This signalled increased politicisation and polarisation. The societies were considered to be a challenge to the constituted authority of the state. From then on, part of the energies of its members went into defence of the right to organise and demonstrate, to occupy public spaces. At the same time lines of social cleavage began to shift. Middle-class members had little common cause with an organisation seen as actually or potentially subversive (unless to control it), and their presence declined.

In 1905 the mutual aid society gave birth to a labour organisation called the *Circolo Avvenire*, which recruited increasing numbers of unskilled workers. The *Circolo* provided an organisational umbrella for 'a building co-operative with 700 members ... an educational association linked with the Popular University of Milan ... a band and choral group, a consumer's co-operative and several irregularly appearing local journals' (Bell 1986: 50). The *Circolo* took over part of the responsibility for running the library, which had more than 3,000 volumes, including a growing range of socialist texts and national and local newspapers. It was also a family circle since, as the local journal commented in 1920,

[t]he old-fashioned family is crumbling. Women are unable to fulfil the domestic responsibility which previously was a basis of family life. One's house? It has ceased to matter. One's children? Working-class parents are unable to watch over them, nor can they assure them sustenance or an education. Both children and parents are made to suffer in these circumstances. (Bell 1986: 50)

To meet their needs it acted as an educational and recreational institution for children.

The *Circolo* was the main forum for the complex and fluid party politics of this period. The Socialist Party (PSI) had been established in 1891, partly in opposition to the largely rural, communal and anarchist organisations which had conducted a campaign of 'spontaneous' violent action against the Italian state since its formation in 1860. The PSI itself had a 'maximum programme' which embraced the full socialisation of the means of production and exchange. But if this was the end, there was considerable disagreement about the means to achieve it – not least over the role of parliamentary democracy. The leaders who made the most radical noises about long-term objectives often advocated the most reformist short-term policies. If there were striking divergences within the Socialist Party (and within anarchism), there were often convergences between activists from different political cultures over revolutionary goals and strategies. Gramsci himself worked with anarcho-syndicalists (whose national union swelled to 800,000 members in 1920) in the factory councils, before becoming convinced of the need for a new kind of revolutionary party.

One crucial factor affecting the ideology and practice of the Socialist Party was the increasingly active role of the church, which was building a coalition which extended from the professional middle class through to peasants, artisans and sections of the new factory labour force. Some of the most salient features of the Socialist Party in Sesto and nationally – its anti-clericalism and its symbolic life – derived from the increasingly bitter hostility to the political activity of the church. There were other political forces in Italian society, but the long-term trend was towards polarisation, and the various coalitions which emerged – whether led by liberals, Catholics or fascists – were dominated by an anti-socialist strategy.

The management practices which had so weakened the position of the skilled workers, and flattened out the terms of employment, had also laid the basis for new and more extensive forms of organisation in the workplace. Craft unions gave way to those based on an industry, and in Sesto the metalworkers union (FIOM) doubled its membership between 1919 and 1920, to cover virtually the entire workforce in the industry. Historians debate the precise sequence and relative predominance of different forms of action in the 'Red Years'; however the evidence suggests that, at Sesto at least, it started as standard union action for higher wages in the period of post-war inflation. The employers responded to the demands with extraordinary intransigence. In August 1920, the employers' negotiator expressed this robust line on the future of negotiations, telling the union chiefs, 'All discussion is useless. The industrialists will not grant any increase at all. Since the end of the war, they have done nothing but drop their pants. We've had enough. Now we're going to start in on you' (Bell 1986: 110).

The employers prepared for a lock-out. On hearing of plans to close the steel works, the workers occupied the Sesto factories, and the political agenda widened. Red flags flew from the chimneys; red guards armed with rifles and grenades patrolled the perimeter; townspeople kept the occupiers stocked with food. Factory councils emerged in the summer of 1920 and attempts were made to maintain some level of industrial production. The issue went beyond whether metalworkers could do their jobs unsupervised, to whether workers could run factories and police the streets. Political relations amongst the workers were not always fraternal. Earlier in the year, when the Turin workers had staged a large and unsuccessful factory occupation, the Milan edition of the socialist newspaper *Avanti* had even refused to print appeals from the Turin strikers (Bell 1986: 109). Milan (though not perhaps Sesto itself) was seen as a stronghold of the reformist wing of the Socialist Party, dominated by craft unions.

In September 1920 the government finally stepped in and forced the re-opening of negotiations between employers and unions, which led to a settlement on wage demands and a promise of parliamentary action to enable worker participation in industry. For the left the decision on the way forward rested on an extraordinarily complex assessment of the power and resources available to each side. If an insurrectionary strategy was adopted, would the various factions of the left stay together, and had enough political work been done to co-ordinate the struggle, above all between the towns and the many radicalised movements in the countryside? How much military force was the government prepared to use, and would the army itself maintain discipline or fragment? The military balance in turn depended on the degree of co-ordination: it was agreed that even the Turin workers, who were the most disciplined and well-armed, would succumb quickly to the army unless action was general and well co-ordinated. You cannot half make an insurrection, or try it out in one city; this was a point of no return. Throughout much of Europe the left was facing the same strategic decision, and this was certainly one of the most dramatic moments in Italian history.

There has been continuous debate about the political possibilities of the 'Red Years' in Italy, and it starts with the question about how such a strategic decision should be made. Gramsci himself did not think that it could be reached through consultation and representative assemblies; it was like war (Levy 1999: 199). In these rare moments of polarisation and uncertainty, the balance of force can shift very quickly: a credible strategy can become suicidal if its opponents have a week to prepare a response. Timing and anticipation are crucial. The Socialist Party met and debated a motion which effectively would have informed the government of the date of the revolution. As a result of the vote, the revolution was adjourned. The majority stepped back, and relinquished their control of the streets and the factories. The wage agreement was hailed as a victory; the factory councils were a learning experience and a re-affirmation of their right to control the workplace, a step on the road to permanent control. A minority, convinced that this had been a revolutionary moment, undone by the lack of preparation and the reformist character of the Socialist Party, joined Gramsci in the split which led to the foundation of the Communist Party of Italy in January 1921.

The industrialists had fewer doubts than the left about what the factory occupations represented: they had just had their productive capital temporarily expropriated and been given a clear message that their presence was superfluous in the coming order of things. In September

1921 Agnelli, in a moment of desperation, had offered to turn Fiat into a workers' co-operative. If they did not take decisive measures immediately, they would lose everything, either through a strengthening of insurrectionary forces or through erosion by parliamentary-backed measures. Throughout the autumn of 1920, money and resources flowed into the fascist organisations, both from the agrarian landlords who had temporarily lost control of their estates, and from those industrialists who had lost faith in democratic solutions to their problems. Within two years, all the workers' organisations at factory and street level were demolished; within five years democracy was dismantled.

CLASS IDENTITY

I now want to step back from this rather dramatic series of events in the years 1900–20, and focus on the formation of political identity. Even by 1914 we seem to be in the presence of a fully-fledged class movement, representing the proletariat in a polarising economic and political landscape. We have a local organisation which united various categories of workers, and was affiliated and responsive to national political associations and parties. Increasingly the actions of the association were co-ordinated and directed by national policy; in the case of the Socialist Party, that of the maximum or ultimate programme. But there is also a paradox. In 1914 the movement had been constructed outside the factories – a set of associations concerned with self-help, education, welfare, consumption and cultural resources. Political solidarity emerged amongst groups of people who continued to experience considerable fragmentation in the workplace. But although at one level all this makes good practical sense, it says nothing about what people do share in the workplace. With this, we move into the core area of working-class identity and interests.

In order to explore this further we need to look more closely at ideas and practices of work, morality and the person. This is a large field and patchily documented; we can easily note dates and variations, but finding a path through it is difficult. I shall take some shortcuts, re-assembling scattered sentences and comments from the historical studies of Sesto San Giovanni and combine them with insights from other sources, including Passerini's very dense analysis based on the oral history of the Turin working class under fascism, and studies of British labour history.

We can find two main conceptions of the relationship between work and the person; two poles which co-exist for long periods, which appear to cross over, and in so doing create a tension at the heart of working-class identity and politics.

The first conception was characteristic of skilled workers, and was anchored around the concept of *mestiere*, which may be partially translated as 'craft' or 'trade'. Men acquired a *mestiere* through a long apprenticeship, and through acquiring the tools of their trade. It involved 'a moral code based on independence, respect and mutual support' (Bell 1986: 44). The dual and contrasting themes of independence and mutuality are important. Mutuality has been touched on already, and derives in part from the need for corporations – organisations which reproduce and control the acquisition of the skills and knowledge which constitute a craft, and hence also from the equality of those who have successfully done so. Equality of status between craftsmen rested on a recognition of hierarchy in the workplace (between the skilled and the unskilled). Independence is multi-stranded; prior to factory employment it meant that there was nobody else in the system who told craftsmen where, when and how to do their job. The craftsmen set their own work rhythm: the technology rarely required them to adapt to the speed of a machine. They also decided how many hours in the day and days in the week they worked. The craftsmen were obviously also responsible for the quality of the work they delivered: it was intrinsic to the whole practice of a *mestiere*.

An indication of the qualities of a *mestiere* comes from the autobiography of Rinaldo Rigola, himself an artisan and later leader of the *Confederazione Generale del Lavoro*:

This cult of the *mestiere* in which the older generation believed has always moved me to reflection. The *mestiere* was in fact an extremely important thing, and the high regard in which it was held is explained by the nature of the precapitalist (that is, pre-factory) society. The *mestiere* was itself a kind of capital; a capital that could be lost and dissipated just like any other, but it was less exposed to such risk than landed or mobile capital. Bankers did not exist who could take it from the possessor. In the *mestiere* there resided both life and independence. (Bell 1986: 240)

Those with a *mestiere* are defined by their work, and define themselves as a person in their work. A whole series of notions about the person and morality are intrinsic to the practice of the trade: skill, honesty, reliability, respect, independence. The work of a craftsman has quality, in the sense that the work done – the product – is a manifestation of the attributes of the person who produces it. By the time of the industrial boom of the early twentieth century, most would have been familiar with wage-labour employment, paid on time rates or on piecework. As we have seen, managers were keen to recruit these craftsmen into the factories; they wanted their skills, but not the rest of the work ethic of independence. In fact there could not be a stronger contrast between the artisanal work ethic and that of Taylorism.

The craftsmen found themselves employed alongside the other main category of workers – the *manovali*. These were 'the hands' (*le mani*), and *manovalenza* became the collective term for a labour force. *Manovali* were hired not for their trade or 'quality', but for their unspecific and undifferentiated working capacity; what mattered was their numbers. They were the masses. They were hired at the factory gate, often simply on the basis of their physical size. The distinction between skilled and unskilled, which we have used as a shorthand, does not convey the distinction between somebody who had a *mestiere* and someone who was a *manovale*. The distinction was not essentially one of greater ability, or capacity to command higher wages; it was between two radically different practices and conceptions of work and the person. These two practices clashed and combined in the formation of working-class identity and politics in the industrial proletariat. Some of the political leaders – including Bordiga, the future head of the Italian Communist Party – held that 'The proletarian is not the producer who exercises his craft, but the individual distinguished from anyone who possesses the instruments of production, and from anyone free from the need to live by selling his own labour' (Williams 1975: 80, 176, 183). In other words those who had a *mestiere*, which included much of the trade-union membership, should be excluded from the ranks of the proletarian movement – because they were property holders, they had the kinds of intellectual and material capital referred to by Rigola, and thus inevitably they had reformist aspirations. We shall return to this issue below.

We have seen that the craft workers provided the core of early political mobilisation in Sesto, and this gives support to the argument that early working-class radicalism owes more to the defensive strategies of those whose lives were doomed by new technology than it does to the oppression of the new masses in the factories. Culhoun (1988), for example, describes a work ethic amongst British craft workers in the nineteenth century very similar to the one outlined for northern Italy, and suggests that they were reactionary radicals, fighting to save an artisan tradition and community life in the face of industrial capitalism; radicalised precisely because no political concessions could save them from their economic fate. The same argument could be made in the Italian context, but with an important qualification. A skilled metalworker, commuting out on a tram from the opulent commercial centre of Milan (the first European city to be electrified) to work in the thundering steel mills of Falck, or to build railway rolling-stock at Breda, was not driven to political mobilisation by the hope of working in a village smithy, shoeing horses. The defensiveness is not anti-modernity; it is about control of the work process and respect for the person.

This political strand endured even under the difficult conditions of fascism. Passerini has an interview with a mechanic describing factory conditions in the 1930s, too long to reproduce here but full of nuances about the practice of management and the ethics of work. Passerini herself (1979: 93–5) comments on a number of themes in the interview which revolve around the shifting balance between autonomy and hierarchy on the shop floor. Since work is associated with ability and honesty it should be self-regulating, hence the strong objections to piecework; hierarchy should be based on merit – founded on, and limited by, work-based competence. I want to draw attention to only one of the points which emerge from Passerini's discussion. The phrase 'it is not my job' runs like a leitmotif through a long section of the interview ('It's not my job, I'm no guard'; 'Your job was to come and tell me'). We can hear the phrase in a number of ways, but I think Passerini is right to stress the moral component. 'It is not my job' can be glossed as: 'I take responsibility for doing the work for which I was trained or hired; to ask me to do other things is derogatory to me as a person.' In the 1970s I heard the same arguments and the same moral anger from an Italian mason objecting to the meddling of his employer: 'If you do not think I am up to the job you should not have hired me. Once you have hired me, keep your nose out of my business.' Basically this is also the work ideology of academics.

The tradition of the skilled artisan gave much of the initial impetus to the workers' movement, in terms of organisation, struggles arising from a distinctive work ethic, and above all a sharp understanding of the alienation of factory employment. As the movement expanded and con-solidated, across occupational categories, and in a national and international arena, the defining features of the workers came increas-ingly to be those of the unskilled manual labourer, precisely those who were not heirs to the craft tradition and work ethic. This is the 'cross-over' and paradox I mentioned earlier. What makes somebody a worker – what is shared by 'the masses' – is an undifferentiated, unspecialised, impersonal labour power, bought in the marketplace and exploited in the workplace. Marx's analysis of capitalist production and the role of the proletariat itself feeds into the political movement and its self-definition.

Hobsbawm (1984: chapter 6) analyses one aspect of this in his discussion of the iconography of working-class movements. He traces the way images of the worker in the period of mass industrialisation became increasingly masculine. On banners and posters there emerged an increasingly fixed repertoire of images: not a craftsman at a lathe, not a woman in a cotton mill, but a man, bare-chested, with rippling muscles, banging heavy metal. Of course the image evokes force, and if its purpose is to drive fear into the heart of the bourgeoisie, it may be more effective than the image of a bespectacled print worker. Hobsbawm suggests that the movement was essentially workerist: 'For most workers, whatever their skill, the criterion of belonging to their class was precisely the performance of manual, physical labour'; and that 'the movement wished to stress precisely its inclusive character' (1984b: 101). We should add that this 'inclusive' icon of the worker remains very specific and essentialised – not just because it is so gendered, but because the labour-power that workers sell is manual, that of the undifferentiated masses. There is no evocation of the skills and knowledge that workers bring to production, nor of the more personalised work ideology documented above. The absence of such references, though understandable, risks collapsing class consciousness into the terms set by the dominant ideology, both in the distinction between manual and mental labour, and in representation of the 'economy' and material interests.

Identities do not of course derive simply from images. I suggested in the Introduction that identity is best understood as a complex narrative, politically constructed, defining boundaries and oppositions. Narratives position collectivities in social processes, in both time and space; they are collective, the work of many minds; they may be more or less stable over time and more or less standardised in their form, depending on the social processes and the organisation forms (such as the Communist Party) which shape them. They are also, in another sense, highly personal. The narrative of the industrial proletariat of Sesto begins, like others, with the new, with a disjuncture. Early political organisers in the town – like Carlo Borromeo, a printer who had earned his livelihood walking or 'tramping' around Europe, and in the process learned a great deal about socialism – knew that they were living in a world of rapidly changing possibilities. Technological developments, such as mass electrification, were changing the character of human labour and the rhythms of life; railways and the internal combustion engine were transforming distance and accelerating the mixing of goods and people. We could say that there was space–time compression. The rapidly accumulating factory force represented a new social reality, created through rupture and disaggregation, exposed to various forms of hardship and dehumanising working conditions.

Ethno-nationalist movements of the period, which often emerged in response to the same modernising processes, tended to create identity narratives around historical continuity, and to create political strategies which dealt with present disruption through the evocation of a harmonious past. Narratives of the working class, especially in this 'unconsolidated' period, defined who 'we' are in terms of who 'we' will become; they are dominated by the future tense. The point is obvious but telling in the names and symbols of the workers' movement: the first political organisation in Sesto was the *Circolo Avvenire* ('the future'); the first local newspaper was *Il Domani* ('tomorrow'); the first socialist newspaper was *Avanti* ('forward'). One of the most popular symbols was the rising sun – a new tomorrow. The sun rose increasingly in the east after 1917. The famous songs – 'The Red Flag', the *Inno dei Lavoratori* – which became anthems of the working-class movement, use only verbs in the future tense.

The processes which were constituting the working class as a collectivity were also creating the base for a new social order. With historical hindsight we may misrepresent the timescales within which the early class movement conceived of its life, assuming that the dislocation and disjunctures of industrialisation would be followed by decades or generations of political growth and consolidation: only at this point would a revolutionary transformation become possible. But the social world of towns like Sesto was doubly new. Even while the industrialists were assembling and disciplining a factory labour force, the workers developed aspirations for a further transformation of production and of society. The ferment of 1920 shows that this was a personal aspiration to be realised in the imminent future, or at least in the lifetimes of the workers – not something for an unknown future generation. Passerini brings out very clearly the tensions that this 'doubly new' reality created in one crucial field: the ideology of work. How was it possible to combine the notion that work is a moral and social duty – that a person derives his or her identity and can realise himself or herself in the work process – with the notion that it is legitimate to strike, to destroy that which has been produced, or sabotage the machinery? (Passerini 1979: 103).

The theme is discussed in relation to the theoretical contribution of the *Ordine Nuovo* group in Turin, which Passerini argues (1979: 95) did not succeed in solving two major problems: first, how to keep the balance between the appeal to contemporary values and the need to overthrow the existing order, and second, how to avoid reducing the concept of producer to that of professional worker, thus forgetting the needs of the increasing masses of the unskilled. At issue here is also the concept of alienation. There is clearly an overlap between Marxist theoretical understandings of alienation (and exploitation), such as those developed by the *Ordine Nuovo* group, and everyday practices of workplace resistance which emerge in the lives of skilled and unskilled workers. The relationship between these strands (which Gramsci himself would call 'ideology' and 'commonsense') raises difficult historical questions of evidence and interpretation.

On another point the historical evidence is more straightforward: the concept of revolution was a central organising feature of the class narrative. As a goal to be realised it embraced a whole range of aspirations: the end of exploitation and of subordination, the end of poverty, the achievement of social justice, a moral reckoning, perhaps the end of history. The concept of revolution was not only a goal, it was a way of knowing where we are, a positioning concept in relation to history, politics and morality. It was normally combined with the metaphor of the road. There was an historicist version of 'the road': historical forces were moving us to the day of revolution, we could not be sure where we were in terms of the distance still to be travelled; the revolution might be around the corner or further off – we could only scrutinise the signs. It was similar to a theory of predestination and was embraced by the maximalist wing of the Socialist Party. For other political currents, there was agency as well as determination: the road had to be constructed. Leaders are those who, through their understanding of events and the example they set, take the working class in a particular direction. There are heroes and martyrs. There are also those who betray the movement: through dividing the working class at crucial historical junctures, or by leading them into a blind alley, or by becoming separated from the masses. To repeat the point: the revolutionary goal provides an historical, political and moral framework for the interpretation of the lessons of history, and of present actions: Will they advance us to the revolution? When a class movement is defeated, and the road forward is blocked, then the narrative which gives shape to an identity also enters into crisis. In the autumn of 1920, with mounting evidence of the failure of the factory occupations to achieve any lasting advance, there was bitterness, fragmentation and mental breakdown (for Gramsci himself, see Williams 1975: 297).

Much of this rich political culture was destroyed by fascism and had to be recreated after 1943, in very different social and economic conditions. Some survived, in the form of workplace resistance and in an underground political culture which Mussolini referred to, with grudging respect, as the generation of the indomitable (*la generazione degli irre-ducibili* – Passerini 1979: 96). At this point we no longer have only a narrative of making, of the new; there is also a tradition – and working-class identity in the sense of being part of a political movement is very frequently acquired as a cultural inheritance.

CONCLUSION

There were many strands in the making of a working-class movement in Sesto San Giovanni. There were the craft workers, with their moral codes based on independence and mutuality, a culture within which a person is defined by and in his or her work. The factory system eroded these skills and values, and the craft workers' struggle within the factories concentrated on maintaining some controls over the production process. They brought to the movement the tradition of mutual aid organisations, largely outside the factory, and a strong sense of the alienation in the workplace which was a result of capitalist re-organisation. They were joined by growing numbers of 'unskilled' workers, hired for their undifferentiated labour-power. Unlike the craft workers, these were men and women whose pre-factory experiences were portrayed as irrelevant to the new environment; they brought to it neither skills nor values nor organisational traditions; just numbers. In terms of a working-class narrative they were people with no past.

Many political movements developed in these industrial districts. In Sesto, in the long run, a revolutionary Socialist (later Communist) Party came to dominate. In its master narrative, a central historical role was assigned to the proletariat. Out of all the social groups experiencing economic and social dislocation from capitalist growth in contemporary Italy – artisans, domestic workers, peasants, share-croppers – only the proletariat held the key to the future. It was the proletariat, sometimes quite narrowly defined as totally propertyless wage-labourers in industry, who would make the revolution. They represented the new phenomenon created by capitalism, and they alone were detached from any of the economic or cultural forms characteristic of earlier modes of production. As a result they would also be immune to bourgeois and reformist political organisations, like craft unions, co-operatives or parliamentary democracy. This was a highly deterministic analysis: a very direct relationship was posited between social being and social consciousness. The proletariat has an historical goal of achieving communism, and since in this historical period it is a small minority of the workforce, it must become a leading class in a hegemonic coalition of forces against capitalism. In the narrative the historical justification for this hegemonic position is that 'the whole labouring class is destined to become like the factory proletariat' (Gramsci, quoted in Williams 1975: 189).

The leaders of this movement created a narrative in which the revolutionary crisis is a total social event, and the role of the new working class, no less than the old artisans, is charged with moral purpose. Gramsci's piece on the Communist Party (4 September 1920, in Williams

1975: 259) delights in the comparison between the modern proletarian movement and the primitive Christian community, as the heart of a new civilisation in which there will be a new 'order of moral life and the life of the sentiments'. In the present moment the sustaining and constant sentiment is solidarity; in the future, with the foundation of a new society, there will have to be other values,

because the enemy to combat and overcome will no longer be outside the proletariat, will no longer be an external physical power which is controllable and limited ... [instead] the dialectic of the class struggle will be interiorised and within every consciousness, the new man, in his every act, will have to combat the 'bourgeois' lying in ambush for him. (Williams 1975: 227)

This and other texts bring out the general character of the crisis and its resolution. The economic crisis of capitalism may be Gramsci's 'Doomsday machine' (Williams 1975: 231), but what we also get from Gramsci is a sense that *the translation of economic processes into a moral drama is essential in generating revolutionary politics.*

I cite these texts because they are available, as they were to the political activists of 1920 in Sesto. It would require a very different kind of historical analysis to assess how influential they were in the construction of a political culture, an analysis which may now be impossible. They do indicate both the tensions and the convergences in the different strands of this class formation, and the links between economic relations, morality and the future tense in the forging of a class identity.

The revolutionary strand is reconstructed in the class narrative in the very different circumstances following the collapse of fascism and the war of liberation after 1943, and remains strong for many decades; but it slowly becomes diluted and muted. In the representation of 'who we are', the weight of tradition, of a highly institutionalised and sometimes very ritualised political culture, comes to take precedence over 'who we will become'. The future tense is of course important in other political narratives of the twentieth century, not least in those visions of national destiny which had such tragic consequences for European history. However, the revolutionary aspirations of this class movement – the struggle to achieve a fundamentally new social order without class divisions – represent a different relationship between past, present and future from that imagined by nationalism. The loss of a sense that the future might be radically different from the present, a theme which had been a central feature of working-class identity – represents also a collapse into economism. The factory system had created a working class which increasingly shared the same conditions of employment. The class movement had in part taken this reality, and the dominant representations of it, as the basis for identity and unity, constructing an understanding of the workers as defined by their economic position in the production process, as undifferentiated masses selling their labour-power. But at the same time the movement had also stressed, in a variety of formulations, that this constituted alienation and was to be transcended – and while that understanding prevailed, workers' interests were never defined in purely economic terms. If the future tense is lost, demands for improvement remain within the dominant discourse of how the economy operates and of the workers' position within it. The demands are material: for higher wages in order to achieve a higher standard of living, since workers after Italy's economic miracle positioned themselves as much in the realms of consumption as of production. The older theme of alienation resurfaces outside mainstream politics, in the 1969 movements of students and factory workers which shook the north Italian cities. The old questions of work and identity were re-opened from a variety of perspectives, at a time when the labour movement appeared locked into corporatist and reformist policies. It was at this time that the 50-year-old writings on the factory councils were republished and briefly regained some resonance.

SOURCES

The first study of a class movement had to deal with a classic urban industrial workforce, and the first third of this chapter draws on Donald Bell's fascinating study of *Sesto San Giovanni* (1986), a town which has mythic status on the Italian left. The rest of the chapter moves out from that in various directions. We have a second English-language study of Sesto, though it concentrates on the fascist period: Willson's (1993) book on the Marelli factory, which pioneered Taylorism. It raises general questions about solidarity and competition in the workplace, and is particularly valuable for its account of women's experience. There are many other studies of industrial conditions in northern Italy between 1900 and 1920, but we need also to note variations: Sesto, Milan itself, and Turin diverge both because of sociological factors (whether labourers are recruited from the countryside or from artisan workshops; the length of their residence in the city) and in terms of political cultures (reformist, revolutionary, anarchist). Gribaudi (1987) takes a different approach to class identity from that of this chapter, but is particularly valuable for the wealth of data on social life in the working-class neighbourhoods in Turin, and on the economic and geographical movement of families over generations.

The study of Sesto raises questions about the relationship between the political culture of 'skilled' and 'unskilled' workers in the labour movement. This is an important theme in British social history, and can be followed up through the work of E.P. Thompson (1963) and subsequent commentaries (Kaye and McClelland 1990) or the work of E. J. Hobsbawm (1984b).

The political events of the Red Years in Italy (1919–21) have been analysed very extensively. A large proportion of the studies focus on Turin, because it had the most radical workers' movement, was the setting for the work of Gramsci and the factory council movement, and because a large number of the historians concerned with this period were involved with the Italian Communist Party. There are many English-language sources, including Clark (1977), Davidson (1982) and Spriano (1975), while more recently Levy (1999) has reconstructed the important but neglected role of anarchists in the political events of the period. Williams (1975) is a wonderful source, partly because it contains as much as anyone needs to know about the complex, interweaving currents and strategic debates on the Italian left in this period. Above all it brings together the national unfolding of political forces and detailed observations on the day-to-day actions of key players, struggling to read the future in a moment when the stakes were tragically high – it reads in places like a political thriller.

Passerini (1979; 1987) is extremely valuable for anybody interested in research using oral history, and I have drawn on her insights into work ethics and class identity in this context. Her contributions go beyond the scope of this chapter, although I shall be returning to some of the issues on identity later. Finally Lumley (1990) is an excellent guide to the political culture of a much later period (1968–78, and especially in Milan). One strand in this richly documented study is the return of revolutionary politics, and although the composition and context for these movements were very different, they do shed light on the renewed interest in the political experiments and aspirations of 1919–21.

3   ANDALUSIA: EVERYONE OR NO ONE

Anarchist leaders in the nineteenth century produced a handful of classic texts, and there are splendid monographs on anarchist politics in various European regions, but there is no body of theoretical work to compare to Marxism, and far fewer scholars keeping alive an understanding of the characteristics, potentialities and limitations of anarchist politics. Marx himself, and then the communist movement, was locked into a bitter struggle with anarchism from the 1860s, and the rivalry led on occasions to the destruction of anarchist movements – as argued by Chomsky 1969 – and a general tendency to occlude or minimise its presence. Such factors are aggravated by features intrinsic to anarchism itself. The refusal to form political parties and contest elections makes it difficult to estimate support, while the movement had a loose and largely unbureaucratised character, which frequently caused it to disappear from public view. When the state dissolves the syndicates and centres, and there are no strikes or marches to disturb the policeman's watch, has the beast itself disappeared? The work of anthropologists and social historians on everyday forms of resistance has provided a better framework for opening up questions about continuity and discontinuity in political action.

'Before 1914 anarchism had been far more of a driving ideology of revolutionary activists than Marxism over large parts of the world' (Hobsbawm 1994: 74), but its influence declined dramatically in the second half of the twentieth century. This long-term decline has frequently been interpreted as a function of broader historical stages: either anarchism itself is generated in societies at an early stage of development, or it represents an early form of politics. Now that anarchism no longer has a monopoly on 'historical failure', this may be a good time to assess its characteristics, including those complex judgements about its primitive or pre-modern features.

Within Europe, the various currents of anarchism and anarcho-syndicalism were strongest in Spain. Anarchism became a major political force in the 1870s, both amongst rural workers in Andalusia and in the urban industrial centres of Catalonia – a fact which immediately casts doubt on any simple correlation between anarchism and 'development'. Its fortunes over the next 70 years fluctuated dramatically, a consequence not of the dynamics of a millenarian religion, but of the extent of state repression. There were periods of heightened activism between 1903–05 – halted by famine; and in the 'Bolshevik' years 1918–20 – ended by military intervention. Insurrectionary activity resumed in the 1930s, including an uprising at Casa Viejas in Andalusia in 1933, which was put down by a notorious army massacre. It was followed by the Civil War of 1936–39, when the Andalusian anarchists, as part of the Republican alliance, were quickly overrun by Franco's troops moving in from North Africa. Over this 70-year period the movement changed substantially, and was continually challenged by a substantial socialist movement; as in Turin or Milan, there was both rivalry and a crossing over between anarchism and other left-wing currents.

There are substantial historical and ethnographic monographs, available in English, which analyse the dynamics of anarchism and its position within a wider Spanish political history. Although these constitute my source material, this chapter does not attempt a synthesis of the findings or the debates. Instead it is a study which reads these sources 'against the grain' to explore and to elaborate themes which emerged in the previous chapter, so that a wider picture of class movements may be built up: questions of work, identity, political organisation, revolution and moral codes. The chapter starts with a general overview of anarchism in Spain, but then goes on to concentrate on one social environment: the great estates in Andalusia. There are four sections. The chapter begins with the revolutionary narrative so that readers gain an impression of the drama of the struggle. This is followed by an account of the organisation of the estates, and then two sections on political action: workplace resistance and revolutionary mobilisation.

ANARCHISM IN SPAIN

A first encounter with almost any of the classic studies of Andalusian anarchism leaves two strong impressions. The first is the level of violence that the Spanish state routinely employed to nullify attempts by the labouring population to improve their conditions. Anarchists also used violence. The more conspiratorial strands within the movement at the end of the nineteenth century used assassination as a strategy to destabilise the state, and anarchists killed large numbers of their opponents after their short-lived victory in 1936. This happened in a context where the state employed a branch of the army – the civil guard – as a disciplined and autonomous 'occupying' force in every town;

where union and political organisations were regularly banned; where mass arrests and torture of suspected activists were common: military columns were sent out to deal with larger-scale disturbances, culminating in a civil war in which one in ten of the population of Andalusia was killed in reprisals after the fighting had ceased. Throughout the Franco regime, and into the 1970s, Spain was the most policed nation in Europe (Foweraker 1989: 182). It sometimes seems extraordinary that anybody raised their head, given the high price paid for activism.

The second aspect which surfaces repeatedly in the studies is the profound immobility of these social forces. It is hard to think of other regions of Europe where there was so desperately little cumulative change over a 100-year period (roughly 1850–1950) in terms of increases in living standards – let alone the redistribution of property or the acquisition of political rights. Mass strikes to increase wages for one harvest, even if successful, were annulled for the next – each year the struggle started again from scratch. After two generations of large-scale mobilisation, in 1905 Andalusia suffered what is generally described as the last major famine in western Europe, with no serious attempt by landlords or the government to alleviate misery:

In the country districts, in spite of all the strikes and insurrections ... [anarchism] has achieved practically nothing. Whether agriculture was booming or slumping, the standard of living of the agricultural labourers in the south of Spain remained practically the same from 1870 to 1936. (Brenan 1960: 157)

Things got worse in the 'hungry years' of the 1940s (Fraser 1973: 71ff.). After Franco's victory, a *pueblo* mayor summarised the view of the elite: 'Since we won the war, you [the labourers] have no longer any rights, only duties' (Gilmore 1980: 136). For a long period, until quite late in the twentieth century, this was not an environment which encouraged faith in the unfolding of material forces leading to progress, or a gradualist approach to social transformation. Over this period there were, of course, changes in political culture and political strategy; nevertheless the temper of anarchist politics is shaped by an enduring concern with morality and justice, and by the bleak alternatives of stasis or the uncertainties of revolutionary violence.

The beginning of Spanish anarchism is attributed to the work of the Italian engineer Giuseppe Fanelli, sent by Bakunin to Spain in 1868. Brenan describes the result in the following terms:

Within the space of less than three months, without knowing a word of Spanish or meeting more than an occasional Spaniard who understood his French or Italian, he had launched a movement that was to endure, with wave-like advances and recessions, for the next seventy years and to affect profoundly the destinies of Spain. (Brenan 1960: 140)

The description of this visit suggests a spark applied to dry tinder, or in the words of one local activist, 'It is just as if someone is in a dark room and the light is turned on' (Mintz 1982: 31). Of course such dramatic results were only possible because there was considerable convergence between existing political aspirations and those of the International Brotherhood.

The anarchist proselytisers developed 'The Idea': that human beings must be free; that all authoritarian relations should be abolished, especially those of the church and the state; that liberty will be achieved through revolutionary acts which destroy those relationships, not by seizing control of the state or trying to reform it. The movement which works to achieve liberty (and here the contrast with communist political practice is most evident) must not recreate the authoritarian relations which it seeks to destroy; participation, whether individually or in local collectivities, must be voluntary and spontaneous. The Idea arrived in a society where the church was strongly identified with the rich, and where, since the liberal land reforms, there had been an increase in rural uprisings against the landlords and the local agencies of the state.

Although most of the message was not new, there was tremendous respect for the form it took. Literacy was highly valued as the key to understanding The Idea, and literacy programmes spread rapidly (Kaplan 1977: 88; Mintz 1982: 85). By 1918 more than 50 towns in Andalusia had libertarian newspapers (Brenan 1960: 179). Learning raised consciousness, and those with the greatest understanding and commitment were 'the men with ideas' – the *obreros conscientes* – amongst whom the artisans of the *pueblo* were strongly represented. Literacy served not just the diffusion of political ideas, but also the acquisition of scientific knowledge, since it was a tenet of anarchist thought that human beings could learn much about themselves and the foundations of society from the study of nature. Science was also seen as a weapon against Catholic superstition. In the larger centres like Jerez, mass secular education, anti-clericalism and women's emancipation were central components of anarchist popular culture from the 1870s (Kaplan 1977, chapter 3), issues which would still be considered radical nearly a century later.

By the end of the First World War – after many unsuccessful revolts, organisational experiments and congressional debates, interspersed with periods of state repression – the dominant political current in Andalusia was anarcho-syndicalism. The *sindicato* (or *centro*) was the main organising institution – it united all those union members who worked in a particular locality. It was a flexible organisation which united people in the workplace, and also more widely in the city or *pueblo*, drawing in artisans, women and the seasonally unemployed. It was a compromise between the two currents which Kaplan identifies so clearly in the political developments of an earlier generation in Jerez: the unionised workplace which mobilised through strike action, and the community of the poor mobilised through local insurrection. For the *sindicato* the main political weapon was the general strike, which brought together the employed and the unemployed – even the shopkeepers and the servants – in a short, complete stoppage. Witnesses to these sudden and total events suggest that different actors may have had different understandings and expectations of them – for some they were a workplace struggle over pay and conditions; for others a revolutionary insurrection.

The syndicates were linked in a loose regional and national federation, the CNT, but this organisation had no full-time employees, no union fees and could pay no strike funds; nor could the federation impose a co-ordinated political strategy. The syndicates of each *pueblo* and city decided voluntarily whether to participate in a campaign. These practices were both a necessity (workers were simply too poor to pay union dues or sustain a long strike campaign) and an intrinsic feature of anarchist ideology. The absence of the kind of centralised command structure developed by the Communist Party made it a very different kind of political movement. It was not based on a theoretical understanding of the historical role of the proletariat, but was open to all those disadvantaged by capitalist development, including artisans and the seasonally unemployed. It also had an open democratic texture, as is revealed in accounts of the Sunday assemblies to debate local affairs (Brenan 1960: 151) – hence there were fewer tensions between the rank-and-file and the leadership, though there were tensions between advocates of reformist and revolutionary practices. The comparative absence of formal organisation was also a survival strategy against recurrent oppression: when the state relaxes its grip, the capacity to mobilise emerges very quickly, whereas a political party needs to select and train its *cadres* in order to re-establish a command structure. (Lenin said that it took 10 to 15 years to develop a leader, and one day to lose one.)

TAKING AN AXE TO THE ROOTS: MORALITY, REVOLUTION AND THE FUTURE TENSE

The revolutionary strategies developed by a proletarian vanguard party required a high degree of centralised co-ordination, as well as a strategic sense of the most effective ways to paralyse a complex, interdependent economic system and triumph over the coercive powers of the state. The general strike, as it developed in Andalusia, was appropriate to a more homogeneous society of estates and *pueblos*, and aimed to forge a kind of 'mechanical solidarity' between them. During the strike each *pueblo* would be sealed off – preventing entry of outsiders, whether strike-breakers or the police – and then shut down, in an attempt to construct a common purpose amongst different occupational groupings. The theory behind it was enunciated in a document by the anarchist federation, the CNT, in 1910:

Given a general strike, if all the workers fold their arms at a given moment, this will result in such a substantial upheaval in the history of the present society of exploiters and exploited, that it will inevitably cause an explosion, a clash between antagonistic forces that are struggling today for their survival; [just] as Earth, if it stopped rotating on its own axis, would then crash into some other celestial body. (Mintz 1982: 25)

One of Mintz's informants, reconstructing the dynamics of the uprising and massacre at Casa Viejas in 1933, had absorbed this understanding:

It is very easy to transform the world. You abolish classes and change the social order. I forget who said it, but it's been said that when all the workers cross their arms and stop working, capitalism will be at an end. What is difficult is to have all the workers cross their arms at the same time – to agree to do it at the same time. The transformation itself is simple. When the workers are in accord, they can change the world at once. (Mintz 1982: 25–6)

One important strand of the revolutionary narrative, and of revolutionary practice in the uprisings and the Civil War, is fuelled by a particular moral vision which relates less to the economic crises of a capitalist system than to the evils of an authoritarian and competitive society. The moral urgency can be heard in other sources. Ascaso was an anarchist leader who sent this farewell message to his comrades, before deportation to West Africa for his part in the Catalan uprising of 1932: 'Poor bourgeoisie that needs to resort to these means in order to survive. It should not surprise us ... no one dies without showing their claws ... Something is breaking down and dying. Its death is our life, our liberation' (Mintz 1982: 144).

A moral purpose, based on a view that there have been centuries of injustice, may give rise to a revolutionary narrative that is detached from any view of maturing economic processes, of progress, or a sense that there are material or spiritual achievements in the bourgeois world which can be carried forward. As a result, in this strand, revolution is associated with destruction, and now is as good a time as any other to make it happen. Back in 1891, 4,000 labourers had marched in Jerez chanting, 'We cannot wait another day – long live Anarchy' (Brenan

1960: 162). Obviously nobody had told them that Andalusians were supposed to prefer *mañana*. The most famous and apocalyptic version of this understanding comes from an anarchist standing with Brenan, watching Malaga burn: 'Yes, they are burning it down. And I tell you – not one stone will be left on another stone – no, not a plant nor even a cabbage will grow there, so that there may be no more wickedness in the world' (Brenan 1960: 189).

This is, of course, not the only strand in anarchism, and if the first stage of revolution was perceived as simple, the second task, that of building, was more complex. José Monroy, another leader of the Casa Viejas uprising, remarked:

We thought that a general strike would triumph. We wanted to live without money, by the interchange of goods. Then it would be another battle. One had to form the organization and then arrange a system for the different organizations to co-operate with each other. (Mintz 1982: 26)

The conviction that money is the root of all evil, and that a future libertarian society would dispense with it, was another enduring strand amongst the *obreros conscientes*. The morality of the post-revolutionary society would be built on sharing, this being in part the projection forward of the values and practices of everyday life amongst field-labourers such as Pepe Pareja: 'If a child has two pieces of bread, he must be told to put one down and not hold one in each hand. That is *egoismo*' (Mintz 1982: 83). These values were carried into adulthood in the fields, where 20 or 30 men stood in a circle and ate gazpacho from one large bowl set on the ground. Another man observed, 'In the *campo* there is one bowl for everyone. If for each spoon you take, he takes three – he eats more bread – it is *egoismo*' (Mintz 1982: 83). The vision of a future society only becomes comprehensible when located in the moral map of the present, with its divisions between rich and poor, or between solidarity (*unión*) and selfishness (*egoismo*).

There was also a growing awareness that spontaneity had generated a series of defeats; that these had a very high human cost; and that experience, leadership and co-ordination were necessary. Here are two other eloquent verdicts on the 1933 Casa Viejas uprising from village activists who participated in it:

We thought that making the revolution was easy. We were young. We were still walking on all fours like a child. We did not have the capacity for revolution. We were ignorant. But in spite of this, we were like caged birds who have no other mission in life than to see if the master will leave the cage-door open. (Manuel Llamas, quoted in Mintz 1982: 265)

Since this town was aroused to revolutionary fervour, it can be said that the villagers believed that everyone else was also alerted. And that being the case, the government could not repress the entire nation. For if an alert like this had taken place throughout the whole country, what would have happened? The forces stationed in each town would not have been enough to suppress a general movement. Don't you think so? That is what happened here. And this little corner of the world remained with its belief. And here there was a massacre. (Pepe Pareja, quoted in Mintz 1982: 265)

THE LATIFUNDIA

We can now go back to examine the kind of society which generated such dramatic political practice, and in particular the shape taken by the rural conflicts. In both liberal and Marxist histories, the opposition between town and country is usually an opposition between progress and backwardness, and the fact that anarchism was strong in the rural economy of Andalusia feeds into the assumption that it represented a primitive kind of politics. But the trajectory of wage-labour in agriculture is generally the reverse of that in industry: it expands rapidly in the last half of the nineteenth century, and thereafter – with technological innovation – declines in favour of family farming. In Andalusia, as throughout southern Europe, the rural proletariat outnumbered the industrial proletariat until the second half of the twentieth century, and since it was generally a radicalised proletariat, rural areas were often a stronghold of the political left.

Agriculture dominated the Andalusian economy before the advent of mass tourism in the 1960s, but there was not a 'rural' population in the sense that the term would be used to describe Tuscany or the Basque country. Andalusia is a region of major cities and smaller nucleated settlements called *pueblos*, with a population ranging from a few hundred to tens of thousands (Corbin:1993: 32). Labourers did not live on the land; indeed those outside the *pueblo* were stigmatised as marginal or rootless. Instead they congregated in the *pueblos* with other occupational groupings – artisans and professionals. The *pueblo* is a 'populated place and a placed population' (Corbin 1993: 72), and this duality is important in understanding anarchist politics.

Outside the *pueblo* lay the estates or *cortijos* which dominated Andalusian agriculture. Large estates, or latifundia, had been dominant in Roman times; since they are no longer found in most other parts of Europe but are still found in Andalusia, there is a temptation to regard this as evidence of unbroken tradition, or as an archaism. In turn, this may suggest a determining role for geography – the presence of the large estates being dictated by soils and climate – thus 'naturalising' a complex social system. In practice, agrarian systems depend on many factors, including technology, markets and the state. Andalusia has areas of successful small farms, while throughout history the maintenance of large estates has required the expenditure of considerable force.

In the dry zones the estates produced wheat and olives, with the main labour input from May to July, and from December to January; wages earned during those periods had to provide an income for the whole year. Struggles over employment contracts were concentrated at harvest time, and were at their sharpest in the years of abundance. The famous periodicity of rural conflict in Andalusia is not explained by hunger and despair; instead, as Aya (1975) demonstrates, the bargaining power of workers was much weaker in the hungry years. During the rest of the year unemployment was very high, though there was some demand for labour in the irrigated zones of the river basins. A small group of permanent employees lived on the *cortijo*, but the main workforce consisted of seasonal gangs of around 25 labourers, living in temporary accommodation on the *cortijo* for a few weeks at a time for the harvests. Women constituted about a third of the workforce, and were employed especially in the harvesting of olives and cotton. Although the *cortijo* was run with wage-labour, it could not be the only framework for political mobilisation, since there was both a seasonal pattern of employment and a lack of stability in the composition of the work gangs.

Andalusia had one of the most polarised agrarian structures in Europe, with 1 per cent of the landowners possessing more than half of the land, mostly in estates larger than 250 hectares (Aya 1975: 18–19; Malefakis 1970: 31). Of the rural population 50 to 60 per cent were landless labourers – close to a million people during the Franco regime – though the total figure started to drop in the 1960s. However, there was no sharp demarcation between the totally propertyless and those who owned small plots of land or worked as artisans, since they were all likely to be employed as labourers during the harvests.

Rural polarisation had been aggravated by the liberal reforms of the nineteenth century, which had sold off huge swathes of church property in 1836–37, sold off communal land in 1854–56, and removed restrictions on the sale of lands held by the nobility (Kaplan 1977: 42). There was a massive transfer of property rights to a new group of wealthy nobles and bourgeois, but not to the peasantry or to the landless. Andalusia had never been an egalitarian society, but these liberal measures sharply reduced the economic options for the poor: long-term tenancy arrangements were replaced by short leases, tenancies in general gave way to wage-labour and communal rights disappeared. The taxation base for municipal governments shifted from land resources to consumption taxes, which hit the poor hardest (Aya 1975: 74–5). Meanwhile the population grew rapidly. Historians date the upsurge of agrarian protest in the second half of the nineteenth century to this rapid expansion of market relations in a rural society, and the polarisation which it created. There was increased poverty but, as Kaplan says (1977: 10), it was the presence of wealth which generated anarchism.

WORKPLACE RESISTANCE

At the risk of some simplification it is possible to describe the character of the Andalusian economy as constant from the consolidation of the liberal reforms through to the 1960s. My account of labour relations on the latifundia will derive mostly from the research of Martinez-Alier in the 1960s, which is itself supported on key points by documentary evidence going back to the early part of the twentieth century.

The organisational problems of the latifundia are those common to labour-intensive capitalist agriculture. Firstly, production in agriculture is seasonal. As a result, either the districts carry a population which is unemployed for a large part of the year; or the local population migrates in search of work in the dead months; or workers are recruited in from other regions during the periods of high demand. These tensions remain until agriculture is 'industrialised', with a very much lower number of labourers, permanently employed as machine operators – a development which triggered the definitive exodus of rural populations and transformed the face of Europe from the 1950s onwards.

The second problem is the control of the labour process. Maintaining the quality and quantity of work output is a problem in all systems of wage-labour, but it is easier if it is structured around a production line, at the end of which materialises some standard product at a constant speed. Setting a group of labourers to hoe a field or to pick olives is more difficult. Either managers try to increase the quantity of work by paying piece rates – which then requires measures to maintain quality – or they try to ensure quality (the elimination of all weeds; the picking of all olives) by paying day rates and must then find ways to maintain work rhythms. Both solutions require a great deal of surveillance, usually in the form of field guards.

Martinez-Alier's research, conducted when all autonomous workers' organisations were outlawed, explores the ways in which *unión* (solidarity) survived both as a value embedded in the moral evaluation of workers, and as a set of work practices contesting the landlords' control. The code of *unión* regulated a variety of issues: not accepting employment for less than the minimum wage – or less than the wage of those already employed; hiring practices; preferential treatment for those most in need. Solidarity often broke down, and individual workers adopted other tactics to deal with 'dull economic compulsion'. The same had been true in less oppressive periods: solidarity always had to be constructed in the face of divisive labour practices and alternative commonsense understandings of human nature and motive. I will concentrate on three strands: the importance of place, the anathema of piecework, and the issue of gender.

The preferential employment of labourers from the *pueblo* was a recurring issue. Landlords often recruited from the mountainous areas and from Portugal, while local labourers demanded that outsiders should only be hired if all those available in the *pueblo* had been taken on. The attempt to exclude outsiders from employment has been interpreted as evidence of the strength of 'parochial' *pueblo* loyalty, undermining wider solidarity. Martinez-Alier argues that labourers did demonstrate solidarity with the seasonal migrants, and local practice was largely a question of maintaining union strength in the face of landlords' attempts to bring in rate-busters and strike-breakers. Even so, we should not underestimate the link between 'the people' and place in labourers' self-conceptions, or in their aspirations for local self-government. Some aspects of this emerge in a petition sent in 1931 to the civil governor of Malaga from a village near Ronda – a document which also offers a rare insight into labourers' conception of their rights to land and its product: that it is in some sense theirs even if they do not own it.

After the harvest ... the employers met and decided to give no work to the labourers of this district. This is no way to live. A *pueblo* cannot support itself in this way indefinitely, seeing workers of other districts come to take away our wages. These fields were forged with the sweat of our brows, and now tenant farmers with more land than they can farm themselves ... come to take away the product of this land that is our mother, and this is done solely to force us to emigrate and leave the house and land of our birth. The village was founded by us less than fifty years ago, if the land produces it is because our labour has made it valuable. We have always looked on these lands as something intimately ours, and though we do not own them at least they give us sustenance ... In our own land we are [now] more like strangers (*forasteros*, those from outside) than are the real strangers. (Abbreviated from Corbin 1993: 40)

The second main area of conflict was over piecework, and the reason becomes clear when we see the dynamics of employment. Bringing in a harvest involves a certain quantity of labour: its organisation turns on the numbers employed and the intensity of their work. There were informal agreements about what constituted normal work rates for specified tasks: a rate that was sustainable over a period of days or weeks, without exhaustion ('without taking too much out of one's skin'). The normal work rate evolved over time and was built into the practice of overseeing work when labourers were employed on time rates. This involved 'compliance' with an agreement: '*Cumplir* sets a customary and obligatory minimum standard of effort and quality' (Martinez-Alier 1971: 179ff.). *Cumplir*, as the adherence to agreed work rates, feeds into the practice of solidarity amongst labourers. It is also part of a generalised view that the poor must work, and that in a just society all will work. Individual refusal of *cumplir* is not acceptable and undermines solidarity. This compliance was withdrawn only in times of heightened political tensions, as in the three 'Bolshevik' years, when a practice of 'slow hands' and sabotage in the workplace supplemented all-out strikes.

Landlords preferred to employ labour on piece rates rather than time rates for the harvests, and were able to impose this employment contract on the labourers, except in rare periods when they lacked sufficient backing from the state. Labourers abhorred piece rates; they represented increased exploitation and the destruction of solidarity and equity. Although over a short period labourers might earn more than on day rates, their total income for the harvest, and hence for the year, was reduced. The result was fewer people, working harder, hence higher unemployment and lower incomes. They also objected to share-cropping arrangements as a disguised form of piecework. Brenan's judgement, made in a time of labour militancy, has been much quoted:

The real struggle on the large estates was over *destajo* (piecework). There was a good reason for this. The landlords could not pay decent wages so long as their labourers did so little work. The labourers would not work harder because by doing so they would increase the already cruel unemployment. The landlords got serf labour – that is bad and unwilling labour – but the labourers did not get the one privilege of serfs, which is maintenance. (1960: 175)

Anarchism generally had a more egalitarian conception of gender relations than the Catholic Church, with its stress on hierarchical relations within marriage. The most committed anarchists formed unions not sanctified by the church: this practice, known as 'free love', connoted not promiscuity but free will and commitment; it 'demanded purity and fidelity without clerical or government interference and control' (Mintz 1982: 91). Understanding domestic relations in Andalusia has been made difficult by the overgeneralised ethnography of honour. Although women constituted a third of the labour force, men were considered to have greater overall responsibility for providing a household income. There were demands that no women should be employed while there were men in the *pueblo* unemployed, to prevent men being left at home 'watching the stew' while the women were out in the fields – a circumstance considered 'ugly' or unnatural (Martinez-Alier 1971: 147). Men also said that there would be more *unión* if they were all bachelors, which indicates not the political apathy of women, but how hard it was to maintain workplace solidarity in the face of immediate household needs and responsibilities – the actual operation of 'dull economic compulsion'. Landlords preferred to employ women for certain agricultural tasks, not because they were more docile (the evidence seems to be the opposite), but because they were paid only 80 per cent of men's wages. In 1963 the Fascist Syndicate equalised pay in order to diffuse some of the tensions around unemployment.

Overall, there is evidence that *unión* survived as a strongly-held set of values, but assessment of its effectiveness needs to be contextualised. In a more open political climate, there is no doubt that these practices had increased income, and even for brief periods denied landlords effective control of their estates (Martinez-Alier 1971: 159). When the state repressed open struggle, *unión* could still have some effect on living standards and create resistance to the most exploitative forms of employment. However, the main achievement in these more difficult periods was the maintenance of forms of equality. Perhaps the most vivid slogan of all those reported in the monographs is *'Todos o ninguno'* – everyone or no one. In the 1920s Diaz del Moral observed that, if in a period of high unemployment a few labourers were taken on, there would quickly follow a strike or demonstration to achieve work for all, and called this the dogma of equality (Martinez-Alier 1971: 149).

REVOLUTIONARY ACTION

Forms of solidarity and resistance were the only kinds of political action possible during Franco's dictatorship. In earlier periods they had existed alongside revolutionary politics, and had involved other social groups. Unemployment was the focal point of social protest – the labourers thought that the land could provide work for all, but that the owners, to increase their profits, did not cultivate as intensively as was possible, and skimped on certain agricultural tasks. They saw 'land without men, and men without land' (Martinez-Alier 1971: 116), and with it extreme social inequality. The solution was the division of the estates: *reparto*, a political rallying cry for a hundred years. Sometimes the demand was for collective control, sometimes for individual plots, but whatever the form, *reparto* would generate work and liberty.

The struggle for *reparto* involved a movement which had from the beginning also included artisans, peasant farmers, domestic servants and others. The fact that the anarchist movement constituted an alliance of different economic actors is sometimes seen as one of the reasons for its weakness (cf. Aya 1975). Artisans and smallholders provided a disproportionate number of the activists. They had more tactical leverage, more resources with which to sustain protest movements than the great mass of desperately poor field-labourers, higher levels of literacy and wider social networks. However, the alliance between these two groups has been described as conflictual: in the strong version of this judgement, 'The class struggle had, in short, been imported into the anarchist movement itself' (Aya 1975: 94). The issue of conflict in this alliance needs careful treatment, as do the shifting connotations of the term 'class'.

Artisans, peasants and labourers had all suffered under the liberal policies of the Spanish state in the late nineteenth century, but the effects had been varied. Land redistribution had made peasant farming more vulnerable; the consolidation of national markets made artisan production of textiles and metal goods uncompetitive in the face of northern industry; the changing tax base hit all small producers. In response they had developed a range of organisations characteristic of artisans in this period – credit unions, co-operatives and mutual aid societies. The difficulties come with the historical judgement, found in both Marxist and liberal interpretations, that one party to the alliance constituted a social group over whom history was about to roll, while the other represented the future. Artisans and peasants have been portrayed as part of a pre-industrial world, fighting a rearguard action to maintain productive forms and values doomed to disappear, while the field-labourers were for better or worse already locked into a more advanced capitalist system. If the diagnosis of conflict in the alliance is predicated on the assumption that there were two classes with divergent historical trajectories, then history itself has proved rather more complicated. Certain forms of artisanal production may have become extinct, but the longer-term trend has not been a simple polarisation between capitalists and the propertyless. Various forms of household production (in agriculture and industry) are again widespread; although admittedly they occupy different parts of the economy and use different technology from those of a century ago, and have complex relations with large-scale industry. The field labourers, on the other hand, turned out to be the dying breed, squeezed out by mechanisation.

We can explore further the issue of class and social division by looking at local conceptions of work and identity. The ethnographic material is patchy, and needs extrapolating from the kinds of questions anthropologists used to ask rural people about attitudes to the land. *Trabajo* means work in the sense of manual labour, and is the basis for a dichotomous view of society. '"We" are manual labourers: the roughness or softness of the hands becomes a sign of belonging to "us" or "them". Hands are the best passport ... hands are the best identity papers' (Martinez-Alier 1971: 210). The objection to landlords is both their effect on unemployment, and that, like the civil guard and all those who stand and watch, they are superfluous to production. The quotations from Martinez-Alier's research in the 1960s represent enduring attitudes, as similar quotations from Pitt-Rivers in the 1950s and Diaz del Moral in the 1920s demonstrate.

However, the connection between the concept of *trabajo* and the issues of work ethic and class identity is not straightforward, for two reasons. First, *trabajo* in Andalusia, like *lavoro* in Tuscany and related concepts in other parts of Europe, has (or had) a dominant meaning in terms of working the land – ploughing, planting, hoeing. Thus an overseer who did some ploughing was considered to be doing work; whereas another man said of himself that, 'He had worked since he was a child. It was only in the last eight years, when he became a shepherd, that he did not work' (Martinez-Alier 1971: 209). Thus labourers' self-definition was built in part around a discourse of manual labour and making the land productive; possibly like other rural people, they considered field-labour primary in the sense that it provided food, a universal necessity, and this influenced relations with labourers outside agriculture. The evidence is fragmentary: certainly for a long period these labourers thought that food was the central necessity, and that virtually everything else was a luxury, if not a vice.

The second, more general, issue is that *trabajo*, at least in this usage, is not the kind of abstract term found in political economy: it makes no reference to the social relations (such as employment) within which work is embedded. When Martinez-Alier investigated work attitudes in the 1960s, what he found was a mixture of three elements, with some tensions between them: the dull compulsion of economic relations (employment contracts and practices); revolutionary aspirations for a society where all would have work and dignity; and the complex patterns of workplace resistance which survived under the fascist corporate state. His research was in fact a major influence on Scott's more celebrated analysis of resistance in *Weapons of the Weak* (1985).

The concept of *trabajo* also raises complex questions about solidarity and opposition. Peasants do manual labour in the fields – but does that make them labourers? What of wage-labourers who do not work in the fields, or artisans and professionals? In those *pueblos* which gravitated around the latifundia, all politics revolved around the opposition between field-labourers and the landlords (Fraser 1973: 47). Other categories were assimilated to one or another side, but the anarchist movement itself does not appear to have been articulated around a generalising concept of work or wage-labour. Evidence for this comes from the circumlocutions to describe different categories of people, such as professionals: 'although they do not work, they also work'; or shepherds; or traders, who belong to 'us' for political purposes, even though they do not work manually.

This is not an argument for linguistic determinism – that such people did not have generalising conceptions of wage-labour or class because they did not have the right words to construct them. *Trabajo* and other terms had been used to develop a theoretical understanding of capitalism, but this does not appear to have become the dominant strand in this local political culture. In Sesto San Giovanni we found a class movement articulated around an identity derived from wage-labour and the historic mission of a proletariat, while other economic groupings were assigned a very subordinate role, and sometimes excluded from the political alliance altogether. Andalusian anarchism was dominated numerically by the field-labourers, but their identity was not constructed around a generalised conception of wage-labour, and anarchism developed a very inclusive political strategy.

The view that there was conflict in the alliance because of divergent historical destinies is oversimplified; nor is there any clear evidence that its participants' visions of the future were incompatible – that there would only be room for one kind of worker in the future. If they had embraced a Marxist-derived movement, that might have been true, since there would have been little room for artisans, peasants and other parts of the petty-bourgeoisie after a successful communist revolution. But in those circumstances even the rural proletariat would have had a rough time: it is hard to see how space would have been found for a million labourers on the state farms of Andalusia. Instead, the arguments come back to a narrower issue of political strategy: not 'What is oppressing us?', nor 'What is our goal?' but 'How do we get there?'

The key issue was the scope and effectiveness of strikes. On the estates, strike action could start around demands over pay and conditions and be settled or defeated in those terms. It might also, however, be incorporated or redirected as part of a general strike, where the anarchist sections brought out labourers and the self-employed, paralysing all local economic and political activity. In the nineteenth century this would have been a prelude to the seizing of municipal government, the burning of tax registers and other property, until the state intervened with military force. In the first part of the twentieth century this range of actions had largely dropped out of the political repertoire: the revolutionary strategy was built around a paralysing general strike and its co-ordination between *pueblos* and the cities. A general strike has an immense capacity to polarise local social relations: it forces all the 'middle strata' – artisans, traders, peasants, overseers, muleteers – to take sides, to identify themselves publicly, through their actions and in ways which might be irreversible, with one or other of the groupings on the economic fault-line of the *pueblo*: labourers and landlords. Rather than simply reflecting existing conflicts, the political divisions of the *pueblo* and the region are constituted through the general strike.

In these circumstances 'reformist' strikes over pay did not preclude hopes for *reparto*, or that events might escalate into a revolutionary situation. But there were tensions within the coalition over the conduct of general strikes, and accusations that 'hotheads' in the movement sometimes weakened it by trying to turn every strike into a insurrec-tionary confrontation. Artisans and peasants, who often formed a majority of the '*obreros conscientes*', had no 'reformist' demands to achieve and had the tactical leverage to engage in more prolonged action than the field-labourers. As a result the activists blamed the collapse of political action on the limited resolve of field-labourers, while the attempt to turn all strike action into a revolutionary wave could jeopardise the possibility of short-term gains for the labourers. In the end there could be no victory for either grouping in the alliance without defeating both the repressive apparatus of the state, concentrated in the towns, and the landlords who were the dominant class in the region. With the military insurrection against the Republican government and the outbreak of civil war in 1936, the tensions in the alliance were submerged in the renewal of direct action against the state, with short-lived seizure of local government, the formation of militias, the burning of churches, and the killing of landlords and clergy.

CONCLUSION

There is something in the temper of anarchism which evokes millennial movements – the term has been used by commentators from Diaz del Moral to Hobsbawm – but it is so freighted with misleading assumptions that it is best avoided. In part, the problems with the term are those which arise whenever a political movement is described as religious since, as Kaplan (1977) and others have remarked, religion creeps into the analysis whenever actions are seen as either irrational or involving strong emotions. The overall argument of this book is that all transfor-matory political movements work through the articulation of identities, generating moral judgements and strong passions – that is, they all have features associated with a stereotypical notion of religion.

In addition, 'millenarianism' evokes a kind of religion which is particularly 'irrational' in its aspirations for a better world, and in its assumptions about how that will be realised, flourishing only amongst backward populations not yet integrated into modernity. Judgements about backwardness or marginality are always embedded in power relations. Even allowing for this, there was nothing particularly backward about Andalusia – it was not isolated from the economic and cultural circuits of the rest of Europe: its agriculture was integrated into a capitalist economic system, its field-labourers were fighting over the same issues of pay and workplace control as their contemporaries in other parts of Europe, its social programme on education, gender and democratic rights was highly 'advanced'. That leaves two other aspects of anarchism which have been seen as irrational: its utopian character and its political strategy.

Anarchism was not a monolithic movement, and was itself only one strand of the left in Andalusia; my intention is not to generalise about the politics of the region, but to examine a political current which was significantly different from those derived directly or indirectly from Marxism. Within that current there was one feature which does seem to echo Christian millenarianism – namely, the fact that certain ideas about time and moral absolutism were combined in political practice.

Above all else, anarchist labourers valued human liberty and its corollary, equality. They were committed to the destruction of all authoritarian social relations, in the domestic and the public sphere, and credited a more advanced form of consciousness to those (the *obreros conscientes*) who had gone the furthest, or were the most intransigent, in realising this ambition. The movement evolved in a society where wealthy landlords dominated landless labourers, an opposition which was polarised and generalised to other social groups by the strategy of the general strike. This polarisation was constructed in terms of the moral qualities of the poor, who worked and shared, and the rich, who were idle and competitive. Revolutionary ambitions were not confined to economic issues, and the polarised moral world, both before and during the Civil War, extended to include the church, the army and all agencies of the state. At the same time the movement operated in an environment which, while not 'backward', did not deliver economic or civic progress to the majority of its inhabitants.

This absence of forward movement contributes, I think, to the moral absolutism of the struggle, and both together contribute a distinctive twist to the revolutionary narrative. First, the view that, from a local perspective, today is as good as any other to make the revolution happen. There are more or less propitious moments in terms of harvests and dearth, but there is no long-term process in the economic domain which increases or decreases its chances of success. Success will be achieved when everybody 'folds their arms' at the same time, which is the recurring problem of co-ordination. Secondly, the belief that an act of revolution is an act of destruction: that in itself it will create the terrain on which existing moral precepts can be enacted. Nothing else from society as presently constituted will be carried forward; nor is there a sense that the future society will develop, in the sense given to development by liberal or Marxist thought: no evolving division of labour, no growth, no money.

The weaknesses of anarchist political strategy have long been argued over, and are the mirror of its strengths. Anarchists could seize *pueblos*, but to be successful they needed to stretch the resources of the state by seizing them simultaneously, and holding the larger centres with their garrisons. Corbin's summary of the problem mirrors the judgements made by the activists of the Casa Viejas uprising:

> Anarchists also experimented with ways of making local insurrections spread in a revolutionary wave, with no success. They were perfectly aware that even very weak state governments had the means to police isolated local disturbances and reverse sporadic local revolutionary successes ... From the perspective of local insurrectionaries, the revolutionary failure was not their actual defeat by an armed force, it was the continued existence of the government which had sent that armed force. Hence the quick collapse of so many local insurrections: the arrival of police or soldiers in itself meant that the revolution had failed and that armed resistance was pointless ... Anarchists had two difficulties with revolutionary waves. They could maintain themselves in power in that place only if anarchists also took power in other places ... However any attempt to organize that spread violated the community autonomy they sought to achieve. Second they could not control provincial capitals and major garrison cities ... revolutionary waves could not be made from the bottom up. (Corbin 1993: 187)

A different argument might be made for anarcho-syndicalism in Catalonia, but it also has to be said that a range of countries was shaken by revolutionary movements in the years after 1918. In Spain and Italy, in Germany and Hungary, communist and anarchist movements developed a variety of political strategies to achieve a revolutionary transformation, but the existing states were stronger than any of them. If we move away from the view that anarchism was the product of backwardness and an 'immature' form of political mobilisation, we can get a better view of its historical importance. For what is striking is how many features have re-emerged in contemporary politics: a stress on equality, and on a wide range of human and civic rights; a preference for small-scale local organisations which network and confederate. There is also a tradition of direct action, though this is not the monopoly of anarchism, or the left. All these aspects, and more, in various combinations and colourations, are core themes in the 'new social movements'.

SOURCES

My own fascination with the political cultures of Andalusia began with reading Martinez-Alier (1971), and I have drawn on that book substantially for the discussion of the latifundia, work relations and the practices of workplace resistance. Fuller accounts of agrarian change can be found in Malefakis (1970), and still-valuable accounts of the political landscape from Brenan (1960). Kaplan (1977) is a vivid and sympathetic account of an earlier period of anarchist history, primarily in an urban environment; Collier (1987) a reminder that not all districts were anarchist. The debate about the modernity and effectiveness of rural anarchism is long and complex: it includes Hobsbawm (1959), which espouses the primitive thesis, Chomsky (1969), which attacks liberal and Marxist scholarship (though in relation to Catalonia), and Aya (1975), which is a very thoughtful and well-documented exploration of revolutionary strategy; he has returned in subsequent publications (1990) to issues of political morality. May (1997) is a very useful overview of the historical literature on anarchism and a reassessment of its relationship to modernity. Corbin (1993) analyses Andalusian society at different levels, and offers a different interpretative frame from this chapter – although I have drawn on its source material and excellent summary of the political strengths and weaknesses of the movement. Pitt-Rivers (1954) and Gilmore (1980) produce contrasting anthropological interpretations of political relations in rural areas, a debate which can be traced again in Martinez-Alier. The section dealing with the activists' own understandings of their political task draws on Mintz (1982): a micro-history of one notorious episode – the Casa Viejas uprising of 1933, reconstructed partly through oral history – it is multi-layered and brilliant. Finally, two books deal with the period since the Civil War: Fraser (1973) is a personal, anecdotal and informative account of social change in one village; Foweraker (1989) is a heavyweight account of political mobilisation in Andalusia and the return of democracy, and a broad-based contribution to political theory.

4   TUSCANY: PEASANTS INTO COMRADES

This chapter widens the range of class movements by examining the mobilisation of a group of workers who were not wage-labourers at all, but share-croppers: the *mezzadri* of central Italy. For more than a decade after the end of the Second World War they struggled to gain control of their farms, in a campaign which dominated political life in these regions and had an enduring impact on the political geography of Italy. It was in many ways a very simple political struggle, generating new forms of solidarity between share-croppers and attacking the linkages between share-croppers and their landlords, both at the level of social practices (such as patronage) and in the ideological representations of the share-cropping system (the *mezzadria*). But in other ways the simplicity is deceptive, and we can learn a good deal about this general phenomenon of class mobilisation from the complex interactions revealed in this case study.

The complexity of this movement derives from at least three issues. The first is intrinsic to the *mezzadria* system, which created a kind of duality in the lives of share-croppers: they were family farmers with considerable (if declining) control over their own working lives, who were at the same time subject to a very overt and personalised regime of surplus extraction. The *mezzadria* itself was ancient and considered the bedrock of social order, especially in Tuscany; but as the twentieth century wore on the system was seen to be in need of some kind of reform. But what would the share-croppers become? There was a variety of proposals, based on different interpretations of the existing system, some seeing share-croppers as in essence family farmers, some seeing them as disguised proletarians. It would also be true to say that competing visions of what constituted justice and an ideal social order influenced the interpretation of the present. Share-croppers mobilised to abolish the existing property relations, but that in itself said little about the kind of society that they wished to create.

The second issue concerns the process of mobilisation and its purposes. Share-croppers formed a majority of the population in the central regions of Italy, but the success of their ambitions for land reform depended on forming alliances with other social categories, as well as action by national government. In post-war Italy there were rival coalitions and competing programmes, but most share-croppers supported a left-wing coalition led by the Communist Party (PCI), and found themselves in a movement dominated by wage-labourers, urban and rural. As we shall see, a great deal of the share-croppers' ambitions, strategy and identity were shaped by the fact that they mobilised within and through the Communist Party, a party which had its own framework for interpreting history and its own strategic objectives.

The third issue is the wider national and international context. The fact that the political organisation for this struggle was provided largely by the Communist Party, and that it took place during the Cold War, had a direct impact on its outcome and on the way the lines of conflict were drawn. This chapter takes forward the previous discussions of identity and the 'road to revolution' by looking at how everyday political struggles are redirected and reshaped by these wider national and international processes.

The organisation of this chapter reflects the need to trace both the mobilisation of a rural population for control over land and their livelihoods, and the wider political context. The long middle section of the chapter deals with the economic and political relations internal to the share-cropping system in Tuscany, and with the various ways this was represented and interpreted. It is based in part on my own ethnography in the region. However, the chapter begins with an account of the Italian Communist Party, covering the strategy and trajectory of the PCI from 1945 through to its dissolution in 1991, and paying particular attention to the issue of revolution. The concluding section returns to the political culture of the period and explores the everyday manifestations of communist identity.

REVOLUTIONARY AND DEMOCRATIC ROADS

The Communist Party of Italy, founded in 1922, was committed to revolutionary action to overthrow the state. It believed (unlike the anarchists) that participation in parliamentary elections was a valuable mobilising strategy, but also that liberal democracy was an incomplete or 'bourgeois' form of democracy. The fascists had a different set of objections to liberal democracy, and it was they who came to power in 1922. Repression and the abolition of democracy occurred regularly in other European countries over the following decades, but the Communist Party in Italy was one of the first to experience this new form of 'populist' reaction. The party drew certain conclusions from this defeat, and afterwards remained locked into its world-view. First, that if the major industrial and agrarian interests were ever seriously challenged in their control over productive resources they would fund armed subversion to regain control. Secondly, that this was not such a complicated operation, because generally the existing repressive organisations of the state (police, army and judiciary) would either stand aside or support reaction.

The PCI survived fascism as a small, clandestine movement, with most of its activists in jail or exile until 1943. The Italian government fell, German and Allied troops invaded, and the newly active communist leaders became the dominant force in the resistance movement in the north and centre of Italy. By 1945 the PCI was the largest political movement, at the head of the Committees of National Liberation which were running northern cities, and with armed partisans at its back. But a second historical event had just boxed off the party's strategic options: Italy fell within the Allied sphere of influence under the agreements negotiated at Yalta. The British and American occupying forces had considerable freedom of manoeuvre to keep it that way; they cut their own deals with supporters of the defeated fascist regime – some of whom were re-enrolled into the security forces – while the partisans were ordered to surrender their guns. Any doubts about what was at stake were dispelled by the tragedy of the Greek Civil War.

Togliatti, the general secretary of the PCI, had returned from exile in the USSR in 1944 and made clear both the implications of Italy's geo-political position (which had been the subject of much intelligence-gathering and speculation within the PCI even before Yalta), and the need to move from being a clandestine vanguard party to a mass party, active throughout Italian society and seeking to become hegemonic within it. Constructing a Communist Party with broadly-based popular support, whose members were not even obliged to subscribe to Marxism, represented a spectacular shift, but it was not accompanied by any change in the practices of democratic centralism in the internal life of the party. A mode of operation which had emerged within a clandestine, scattered, insurrectionary movement, involving top-down decision-making, insistence on unity and intolerance of dissent, continued virtually unchanged within the new party. In the circumstances it is not surprising that there were confusions and disagreements over PCI strategy; about what constituted revolution and how it happened. The PCI was thought to practice '*doppiezza*', a double strategy whereby an apparent commitment to parliamentary democracy served as a cover for insurrectionary ambitions. The accusation was used by the party's opponents as a reason why it was unfit for government; but for a long time this was also a view held within the party itself, both amongst local activists and some of the leadership. We can deal in turn with each of the roots of this 'double' strategy.

Soviet communism had immense importance as a symbol and an example within the PCI , reinforced by the role of the USSR in defeating fascism, and by its economic and technological successes: in the 1960s many economists still believed that the USSR would overtake the West by the end of the century. The view that revolution necessarily involved insurrection was widespread amongst the Italian rank-and-file, and many partisans had not wanted to surrender their guns. When an attempt was made on Togliatti's life in 1948 there were spontaneous uprisings throughout north and central Italy: southern Tuscany was one place where ex-partisans dug up their guns, cut off communications and attacked the police. Even in the 1970s and 1980s, lacerations were caused within the PCI by the debates which concluded that neither the October Revolution nor the Soviet experience constituted valid models for contemporary Italy.

This understanding of revolution-as-insurrection was reinforced by the memory of the fascist rupture with democratic politics. In the 1960s there were more or less serious preparations for military coups; in the 1970s, supposedly a period of international *détente*, the Cold War impacted very heavily on Italian politics. The decade began with the launch of a strategic review by PCI General Secretary Berlinguer, entitled 'Reflections on the facts of Chile', prompted by the military defeat of a democratically-elected Marxist government. Although the extra-parliamentary left concluded that the lesson of Chile was the need for the proletariat to be armed in a revolutionary conjuncture, Berlinguer argued for widening the party's alliance to include 'Catholic' forces. As the decade proceeded, its electoral strength grew and the PCI came closer to government, there was increasing concern about threats to that strategy from what the press began calling, in coded terms, 'parallel' or 'deviated' agencies of the state. There was a series of killings by the Red Brigades, and bombing campaigns against civilians which were the responsibility of more obscure political forces. The decade ended with the assassination of the Christian Democrat leader Aldo Moro at the hands of the Red Brigades, and with it an end to any possibility that the PCI would enter the government. Investigations since have shed some light on the shadowy organisations of the period, a labyrinthine network of old fascists, young terrorists, Masonic lodges, army generals, secret service agents and cabinet ministers. There were links between these networks and military agencies which were part of an old anti-communist CIA-backed 'stay behind' operation called Operation Gladio.

In the case of Moro's death there are still questions about who did what and for whom, but there can be little doubt that the PCI's parliamentary strategy was opposed by some very undemocratic forces.

We can now turn to the 'non-insurrectionary' strategy first outlined by Togliatti in the closing stages of the war. The creation of a mass party was immediately very successful: 1.8 million members were recruited by 1946, which grew to more than two million a few years later, organised in both territorial sections and workplace cells. The elections in 1945 were the first in Italy conducted with universal adult suffrage, and in the following years the PCI became the second-largest party after the Christian Democrats, and allied with the socialists in a Popular Front. But the mass party was not just conceived as an electoral machine, and the leaders did not imagine that putting together an alliance that won 51 per cent of the vote would usher in socialism. Nor did they think politics was confined to the parliamentary arena: the triumph of socialism would depend on long-term transformations within what was termed 'civil society'. So the mass party was also part of a strategy which owed much to the work of Gramsci, and aimed to make the working class the 'directing class' in a range of social contexts. These flanking organisations of unions, co-operatives, women's groups, cultural and leisure circles constituted a 'communist world' which broadened the political culture of the party and created a wider consensus for its programme.

This mass party was, in the early post-war decades, still linked to a revolutionary strategy, but one where revolution was conceived as a process rather than an event. This could be presented within a historical time-frame of the inevitable and irreversible defeat of capitalism, in orthodox Marxist–Leninist mode; but it also implied a break with the October Revolution as a model, and an allowance that there might be multiple roads to socialism. If the revolution was a process, then there would be a period of lengthy transition, and no clear moment when you could say a qualitative change had occurred (see Togliatti in Sassoon 1981: 138). This also implied identifying a programme of strategic reforms – strategic in the sense that they constituted crucial steps in a long-term strategy of socialist transformation, in fields such as workers' democracy and reform of the state. What were the reforms that made a difference?

The PCI could point to its key role in the creation of a republic out of the fascist regime, and in the framing of a new constitution; but after two decades of post-war political stability and economic growth, it became hard to maintain the momentum. Attachment to revolution-as-insurrection had declined amongst rank-and-file activists, but even with 'revolution-as-process', gaps began to open up between ideology and practice. The PCI had a highly disciplined and centralised structure, and talked constantly about socialism and revolution, but other aspects of its behaviour were similar to the European social democrats, who had never aspired to overthrow capitalism. The judgement is perhaps over-simplified, and Sassoon's formulation of the tension (Sassoon 1981: 235–47) is helpful, bringing us back to the time-scales of political action: the strategic problem for the PCI was that revolution-as-process would collapse back into a two-stage operation – before and after. On the one hand there was preparation for winning, building an electoral coalition; and on the other victory itself, conceived as taking over the 'command room' of the Italian state and controlling the levers of power.

By the 1970s the PCI had constructed a very powerful apparatus and had indeed achieved great electoral success. It had widened its social base beyond the industrial and agricultural proletariat to include many white-collar professionals and small business people, and had also begun to conceive of its base of support not just in class terms, but in relation to processes outside the sphere of production. In the 1970s women and students joined the list of social categories that the PCI claimed to represent. It ran a swathe of regions in the 'red belt', and formed the municipal government of most of the country's great cities. In many small towns and urban peripheries a person's social relations were lived almost entirely under the umbrella of communist-led organisations. Yet with all this consensus and such organisational resources, the PCI seemed cautious in pursuing strategic reforms, and – from Togliatti to Berlinguer – continued to believe that alliances with the 'Catholic' middle class were possible. At the grass-roots level enormous energy was devoted to the arithmetic of elections, and the progress of the party seemed to be measured in percentage points – not by the creation of political hegemony within civil society or in the realisation of a programme, though there were some exceptions in local government. The key to electoral success was identified in two themes: the democratic credentials of the party, and its reputation for 'clean hands' in running the local administration. Both were important, but seemed designed to reassure the electorate that the PCI was 'safe'. The party was also cautious in the civil rights field, putting its huge resources behind various progressive causes, but always after these had been opened up by other groups.

Some of the loss of momentum can be attributed to the failure of the PCI to anticipate the directions in which Italian society was moving. It underestimated the dynamics of Italian capitalism in the years of the 'miracle', tried to 'solve' the *mezzadria* in the core 'red belt' while the countryside emptied, and lived in awe of the Catholic Church even as its dominance over popular culture and morality were eroding. It gave the impression of intellectual inertia even when it was innovating, and here the political culture of the Cold War played its part. The routines of Marxist-Leninism did not allow the easy acknowledgement of contributions and insights from outside that tradition, and demanded that each change in party strategy be presented as a continuity – a development of the work of the ancestors.

There are many other reasons why the PCI failed to live up to the ambitions it set itself, and not all of its failures were avoidable, as should be clear from the comments on the Cold War. There are studies tracing the evolution of PCI national strategy, but judgements about what the party achieved and was attempting to achieve can also come from evidence about what was actually happening at a grass-roots level; and here regional variations intervene. The PCI often had a more radical edge where it was an embattled minority (in the Veneto, or parts of the south) than when it formed a majority government. In Emilia-Romagna it co-ordinated a very strong co-operative movement and had an unusual record of innovation, not least in forms of local democracy.

The next section will concentrate on Tuscany, where the PCI became dominant through its mobilisation of the share-croppers. It will use a variety of materials, including my own ethnography, to trace the way a political movement emerged, and how it was shaped both by economic tensions intrinsic to the system and by the competing political cultures within which its demands were articulated. This movement was not urban or proletarian: in a sense it could have gone 'either way', and there are paradoxes in the eventual outcome. The concluding part of the chapter will turn to the political culture of the PCI in Tuscany after 1945, concentrating on collective identities and narratives in the long shadow of the Cold War.

THE *MEZZADRIA*

Tuscany was for centuries a remarkably stable society, built around a hierarchical division between the towns, with their landowning households and commercial activities, and a much larger rural population of share-cropping farmers. These share-croppers were not landless labourers like those of Andalusia or Puglia, nor classic land-owning 'peasants' like those of the Veneto or the Basque country, with their more egalitarian social forms. Landlords owned the farm, the farmhouse (*casa colonica*) and half the working capital, whilst the tenant (*mezzadro*) and his family provided the other half of the working capital and all the labour, with the product divided equally between them. Some landlords owned just two or three farms, others owned great estates with 40 or 50 farms. The farmers (*contadini*) lived outside the town walls in the countryside, a pattern quite unlike the urban conglomerations of the Italian south. Prior to industrialisation the *mezzadria* was the key institution shaping the relationship between town and country – spatially, economically and politically.

There were once millions of share-croppers in north and central Italy, but it had always been seen as a rather unusual system, not fitting the main categories of political economy. British and French observers studied the system in the eighteenth century, as did the Tuscan landlords' agrarian association. Was this system the best way of optimising production? Why was it so stable? Was there a trade-off between productivity and social peace? At the end of the nineteenth century socialist leaders also scrutinised this rural system, and concluded that share-croppers would be hard to mobilise in a revolutionary cause. Then, in the first decade of the twentieth century, a collective movement challenging the landlords began to stir, and it exploded in the 'red years' after the First World War. Central Italian landlords were a key component of the fascist alliance, and they suppressed the rural movement with the same violence used to regain control of the factories and the streets. Fascist intellectuals also studied the share-cropping system, and the regime adopted it as a model of corporate relations between capital and labour, extending it to new areas in Italy. At the end of the Second World War the demand for reform re-emerged and had a lasting impact on the country's political geography, with the share-cropping regions of central Italy becoming the 'red belt', where the left dominated.

These farming households were composed of joint families numbering up to 30 members, though ten was more normal. An intensive farming system deployed a complex division of labour, by age and gender, under the authority of the *capoccia,* the male head of household. The *capoccia* signed the annual contract, pledging the labour of the entire household to the farm, and was responsible for delivering information on stocks for the accounts, as well as often being the only one allowed to attend markets. The contract of a family was renewable – some share-cropping households stayed on the same farm for generations and centuries – but they could also be moved between farms on the estate, or evicted at the end of the year for malpractice.

Some features of this system replicate the kind of peasantry which Marx described as a 'sack of potatoes': households engaged in intensive and diversified production to provide their subsistence needs, living in economic isolation from their neighbours. The difference was that the share-croppers owned neither their farms nor their houses, and had one half of everything they produced taken from them. The form taken by this surplus extraction is central to the economics of the system and its discourses of power. Landlords did not have to worry as much about the quantity or the quality of the work done as they would in a wage-labour system, since normally it was in the interest of *mezzadri* to harvest as much as possible and increase the productivity of the farm. Instead landlords needed eternal vigilance to obtain their half of the produce. This was the old management problem of the *mezzadria*, and in fact we have some very vivid descriptions of it from the thirteenth century. Peasants might look dumb and behave deferentially, but within their own world they were reckoned to be immensely cunning. These same accounts recommended techniques to prevent cheating, the best times to do the rounds, and methods of checking. On one estate I studied, the landlord had even built a watch-tower for surveillance.

The share-croppers' life revolved around the farm and household needs, but the system as a whole had always been part of a wider economy, providing food and raw materials for the towns. In the nineteenth century the market orientation increased within the *mezzadria*, bringing investment in new crops and technology, implemented through an extension of the estate system. It was the estate, with a manager and an administrative centre (the *fattoria*) that purchased the new industrial inputs to farming and stored threshing machines or bulk wine-making facilities. In these circumstances the estate, rather than the farm, became the true unit of production: the landlord sold in bulk the product of many farms, and he alone owned the essential machinery and facilities. Where centralised estates developed, the share-croppers were subject to increased accounting processes and a substantial reduction in their autonomy over farming decisions.

In the period between Italian unification in 1870 and 1945, there was increasing tension within the *mezzadria* between the subsistence logic of the farm and the market orientation of the estate. The balance between the two was the result of a series of strategic decisions made by Tuscan landlords, and informed by studies and debates within their agrarian association. There was a variety of considerations. First there were questions about productivity and the market. The classic *mezzadria* began to appear backward in its technology, perpetuating low levels of productivity and very unresponsive to new markets. In the Lombardy plain, where commercial pressures were felt more strongly, many landlords had responded by switching to specialist production with wage-labour. They created a rural proletariat which, by the end of the nineteenth century, was becoming active under the direction of the rural leagues and the Socialist Party. In Tuscany they noted the lesson: there was a trade-off between productivity and rural unrest. The *mezzadria* was kept in place, and less disruptive ways of introducing market relations were found through the estate system and the selective use of wage-labour. Thus, amongst the more entrepreneurial landlords, commercial life did quicken within the *mezzadria*. However, a combination of inertia and a deliberate strategy by the more politically reflexive landlords kept the share-croppers themselves locked into a subsistence regime. This was because social control was the second part of the landlords' equations.

A Tuscan farmer recounted to me a conversation he had once heard between two landlords. They were discussing the recurring problem of unruly share-croppers, and one of them said, 'Good *mezzadri* need good health, much work, and no money', a phrase which encapsulates the political economy of the period. Landlords of course wanted their half of the product, but they were also alert to the social processes which threatened the rural hierarchy. They did not want any of their farming families having a moment's idleness; nor did they want them near the bars or the market, meeting people. If the share-croppers had money they would spend it, and this would mean, firstly, time subtracted from farmwork, secondly the development of vices (smoking, drinking, gambling), and thirdly meeting people – especially in bars and in the towns, which were a breeding ground for unwholesome political ideas. Even the first stirring of commercial life outside the city was seen as threatening. The Marchese Incontri in 1925 lamented that 'the patriarchal calm of the Tuscan countryside had been spoilt by the bicycle and the circulation of newspapers, which had irremediably broken the isolation of the peasants, thus destroying their characteristic ingenuousness, docility, diligence and parsimony' (Pazzagli 1979: 109).

In the first half of the twentieth century there were many political discourses defining the *mezzadria* in relation to society and history. The oldest and dominant representation was constructed in hierarchical and organic terms: the landlords were the head of the social body, they directed society – they even grew their little fingernails long to show they did no manual labour, and demanded respect from those who did. Landlords inhabited the *città*, and were judged on their civilised values and lifestyle; share-croppers, *contadini*, were judged on their capacity for work. The landlords had the power to control their dress, their consumption habits, and even when they married. *Contadini* had to know their place in the world, which was on the farm; when they fulfilled the obligations of their social position they were invisible to the urban world, and when they were visible it was a sign of vice or transgression.

These organic models of society, and the practices which embodied them, were still detectable in 1970 when I first began fieldwork in Tuscany. You could find plenty of witnesses to the traditional forms of deference (including in one case the obligation to kneel when the landlord arrived on the threshing floor); and you could interview aristocrats who regretted the passing of these ways. However, in the post-war period a second, more liberal, understanding of the *mezzadria* gained ground amongst the system's supporters. In this representation, share-cropping was based on a partnership between capital and labour, with the risks and the benefits shared equitably between the two. Tuscany was fortunate to have developed such an harmonious system, and it was folly to abandon it or tinker with the equal division of the benefits. This more 'market-oriented' discourse, resting on notions of voluntary partnership and equality, is obviously incompatible with organic hierarchies. Nevertheless the two discourses co-existed in the fascist and the post-war period, being elaborated in different social contexts.

Of the two, the hierarchical version made most sense to *contadini*, because it corresponded to their experience of living in an isolated world, with its daily forms of deference and absence of civic rights. This did not mean that share-croppers considered the system just, and we know that they resisted its exactions, but at a certain point they began mobilising in new ways. Historians have analysed the factors producing this political radicalisation, suggesting one which was 'structural' and internal to rural social relations, and another which was more 'conjunctural' and external. The 'structural' factor was the commercialisation of the *mezzadria*, and the subtle but important changes it wrought in landlord–tenant relations. New farming practices meant that the old contracts had to be modified, and were contested. In the past share-croppers had been part owners of the working capital (oxen and implements); now that landlords started to own all the important equipment, the balance between labour and capital shifted, and struggles began over costs and returns. Technological changes increased landlord control over the labour process, and the household increasingly obeyed a daily round of orders – a situation they shared with all the other share-croppers on the estate, strengthening horizontal ties in an environment previously dominated by relations of patronage.

The 'conjunctural' impetus to radicalisation was war, since the two major mobilisations in 1919 and 1945 both followed the demobilisation of armies largely conscripted from the rural population. Between 1915 and 1918, war widened the experience of a whole generation by taking them off their farms for the first time, and ended with promises of land and justice from the government. The share-croppers became major actors in the Red Years from 1919 to 1921. They elected socialist (and populist) councils, demonstrated for the abolition of the *mezzadria*, and achieved significant improvements in their contracts. The fascist reaction reversed the improvements and froze all overt political action. Between

1943 and 1945 war moved through central Italy, totally disrupting the rural economy. Since the *mezzadria* had become so completely identified with the fascist regime, defeat seriously undermined the legitimacy of the landlords, while the PCI, which had been the major organisational force in the resistance, gained legitimacy.

The mobilisation of the share-croppers in 1919, and again in 1945, are major events in Italian political history, and at one level represent a straightforward class polarisation. On one side there were land-owners: when first seriously threatened they backed a reactionary government; and when fascism collapsed they defended themselves by calling in the police to deal with every demonstration, and by endless use of the law to deal with breaches of contract. On the other side we find the *mezzadri* mobilising around an escalating series of demands, starting with the division of the farms' product. The *mezzadria* was built around a very visible and personalised form of surplus extraction, and contesting it focused on two aspects: the customary tribute or *regalie* taken to the landlord's house, and the harvest, which was the focus for strikes on the threshing floor. Women mobilised on all these issues, but in addition wanted contractural recognition as full working members of the farm (*unità di lavoro*).

The *mezzadri* wanted to wrest from the landlords first greater control over estate management, then a greater share of the product; in the end they challenged existing property rights, the foundations on which the *mezzadria* was built. From the accounts of activists in the late 1940s and 1950s this was not an easy step, both because of caution in PCI strategy, and because of the share-croppers' deference. Although private acts of appropriation and disrespect might exist, most share-croppers had been inculcated with pervasive habits of public subservience, and even in the worst of times for many people the landlords' rights to the land retained some kind of legitimacy. An activist, describing the problem of political mobilisation in this environment, used a phrase which he said came from the Spanish anarchists: 'We must create men who cast shadows.'

Those working on the farm saw little that was positive in the towns: just landlords who took away half the food, state officials who did the landlords' bidding or conscripted men into the army, and a crew of other parasites – including the priesthood – who lived up on the hill at their expense. But they took their struggle to the towns, since that was where the power lay, and where their own force would become visible. However, the small provincial centres of Tuscany were a hostile environment dominated by the landlords and the church, extolling an ideology of localism (*campanilismo*), hierarchy and organic unity. The left contested every occasion and every organisation which articulated this model of local unity, and adopted forms of political practice which marked out alternative constructions of society. The model of boss and labourer working quietly together in harmony was counteracted by the two major forms of political action: the strike and the demonstration. From the 1948 uprising, through the major battles over contracts, to the later anti-fascist demonstration, strikes and demonstrations were usually combined, and represented moments of rupture. The country invaded the town, all economic activity was suspended, and the vertical relations between landlord and share-cropper gave way to horizontal relations between people who assembled from all over the province, and who defined themselves around a class, not a local identity.

Yet though the lines of division were simple, the struggle itself was complicated by ambivalence about what the share-croppers actually wanted to achieve and who they would ally with to achieve it. Here we encounter two further representations of the *mezzadria*, this time amongst those who wanted to reform it. Were they peasants *manqué*, eager to take their natural place in a society of property-owning family enterprises? Or were they 'semi-proletarians', whose exploitation made them allies of wage-labourers in the building of socialism? Each vision of a future society led to a different interpretation of the present, while each reforming project was shaped by the increasingly polarised political cultures within Italy as a whole.

Property-owning households were central to the social doctrine which the church had developed in the late nineteenth century in response to liberal and Marxist materialism, and which was an ideological pillar of the Popular Party and the 'white leagues' which developed between 1918 and 1922. These 'white leagues', strongest in the north, had conducted an aggressive campaign against landlords and in favour of property-owning small farmers. The programme re-emerged in the governing Christian-Democrat Party (DC) after 1945. The Italian government enacted a partial land-reform programme, expropriating landlords in various southern provinces and in the Maremma (the coastal plain between Rome and Livorno). Using credit made available by the government, others were able to buy out property rights where the landlord was prepared to sell. In this way the 'anomaly' of the *mezzadria* was resolved through the creation of family farmers by the ruling Catholic Party – though it was always selective, and needed buttressing by a paternalistic reform agency.

In the 'Red Years' from 1919 to 1921 the left had generated a different programme – a Bolshevik model of collectivising the land at the estate level. There was little trace of this programme when mobilisation resumed in 1945, though one of the instruments created in the struggle would have facilitated this transformation. On many estates households elected a *commissione di fattoria*, (a kind of rural 'factory council') to negotiate with the landlord. These could have been an instrument for collective management, or even collective ownership; instead they were dissolved within a few years. The Communist Party, in alliance with the socialists, won control of most local councils in central Italy. Its strategy was based on the building of a broad anti-capitalist alliance in order to take control of national government. The share-croppers were an important part of that alliance, since they were clearly an exploited category of workers, and the party's own analysis had identified ways in which the capitalisation of agriculture had moved the form of exploitation closer to that of the proletariat. However, they remained an anomalous category within that alliance. A speech given at Siena in 1956 by Emilio Sereni, the Communist Party's major theoretician on rural society, reveals how this anomaly was conceived, and how a longer historical time-frame would allow its resolution:

The principal task on the eve of the overthrow of capitalism is to create in the countryside the alliance most favourable for a socialist transformation ... It is clear that for us in this historical context, in this historical phase, it is essential to have a bloc made up of all the working masses in the countryside. This must include those strata of the peasantry who certainly have elements of a capitalist orientation, but who are in opposition, like the rest of the peasantry, to monopoly, to the great landed interests ... It is evident that the aspiration of peasants when it is not illuminated by Marxist doctrine, is a petit-bourgeois aspiration, which through explanation we can direct onto the right road. The road must be the transformation of the share-cropper into the owner of the land, because this is what can be realised in the present phase of society. (In Bonifazi 1979: 86)

The general strategy was based on a broad economic category – all the working masses – and the realisation of a general interest – the construction of socialism. At a national level the PCI had difficulties holding together this 'bloc', whose component parts had a variety of more specific interests and were evolving according to very different rhythms. It supported the share-croppers' demand to own their own farms, but only as an intermediate objective on the road to socialism – and with the very important proviso that its realisation should in no way threaten the unity of the wider alliance on which the attainment of socialism depended. It was in this context that the drama unfolded. The PCI campaigned for a comprehensive land reform programme which would deliver individuated land rights, and found itself in competition with the Christian Democrat government, which also favoured the transformation of share-croppers into family farmers – but selectively, through a piecemeal reform programme and credit schemes. The PCI, with its slogan 'The land is conquered, not bought', was hostile to the initiatives of both the government and the share-croppers themselves, which threatened the solidarity of the movement, especially in the Italian south. But some share-croppers paid a price for this hostility, which emerges in this criticism of PCI strategy from one of those directly involved:

> The possibility of buying the land was considered only in the face of an attack from the landlords, when there was the danger of *mezzadri* and labourers being evicted, and even then only through co-operative forms. The attraction of land-ownership to individuals was under-estimated, above all in areas where a spirit of co-operation and an awareness of collective work was less advanced, despite the presence of large estates. Above all we must note the failure to take the initiative in buying land at a moment when the battle over the contract had already convinced many large landowners that it was impossible to beat the strong peasant organisations. So they decided to sell up. This new battle-ground was not immediately identified by the *Federmezzadri*, to the contrary it gave the initiative to the landlords and to the break-away Catholic organisations. (Bonifazi 1979: 87)

This specific criticism of PCI policy, like the preceding quotation from Sereni, raises wider questions. Take away the revolutionary transformation from this political narrative, concentrate on the 'short-term' question of which political force was best placed to deliver them land reform, and the share-croppers' choices start to look rather different. It brings out the way in which local struggles are transformed by national class politics and channelled into a wider project which incorporates them in a reforming or revolutionary vision of the whole of society. Of course, if 1956 had been the 'eve of the collapse of capitalism', we would be dealing with a very different history. Instead we find that, while family farming emerged in the reform areas, elsewhere there was a massive rural exodus, leaving empty estates which would eventually be run with wage-labour.

The PCI did not make much progress towards a socialist transformation, in Tuscany or anywhere else. Living standards certainly rose, though there remained striking disparities of wealth, especially in the wine districts. The PCI provided a political home for a broad spectrum of social groups, not least the ex-share-croppers who continued to support it after becoming owners of small businesses, farmers, builders, shopkeepers and hoteliers. It is here that the failures of a process of strategic reforms become evident. Tuscany did not develop the extensive co-operative movement found in Emilia, or other ways of linking small producers either in horizontal chains with each other, or in vertical chains with consumers. The small-business sector became very defensive in its political reflexes, often highly dependent on state subsidies and on benign tolerance of tax evasion, and was both vulnerable to wider markets and extensively penetrated by larger industrial interests.

In the 1980s I encountered a group of farmers, all PCI members, who had founded a cattle-rearing co-operative, dependent on subsidies from the Tuscan region. They bought most of their supplies from an international company selling animal feed, and were servicing substantial loans, obtained at very high fixed interest rates from one of Italy's largest banks. Their main sales outlet was a branch of the communist-run co-operative movement, which then precipitated a financial crisis when it switched to buying imported meat. The Tuscan region pulled the plug on the farmers' co-operative. Soon afterwards, local shops in the consumers' co-operative also shut, to be replaced by a branch of the national chain owned by Silvio Berlusconi. This was not a secure environment for small business; it offered few controls over the economic context in which they operated, and little resistance to the growing concentration of resources in the market-place.

THE FORMATION OF COMMUNIST PARTY IDENTITY

The PCI put down deep roots in the Tuscan provinces after 1944, and established very stable forms of political allegiance, while the society of which it was a part changed out of all recognition. Share-croppers, the millennial backbone of Tuscan society, had gone by the 1970s. Some of the migrants went north to Turin and Milan, while the majority moved to towns in the region. Many of their children, the first generation to have access to secondary education, then moved through the universities in the tumult of the 1960s and into the professions. The social base of the Communist Party in Tuscany widened enormously, not just because it went looking for support amongst the middle classes, but because it maintained the allegiance of a million rural people as they moved, geographically and socially, within Italian society. This stability reflects the strong identification with the party which derived from the political programme already discussed, but also from at least two other factors: the social forms and values the PCI developed, and the social context of the Cold War.

The party generated comradeship – a deeply-rooted practice of solidarity and inclusion, offering collective support not just in the rather individuated work relations between share-croppers and their landlords, but in the neighbourhood sections, in the *feste* and other leisure activities.There were symbols of belonging in all the social contexts where the PCI operated. Comradeship was based on an ethic of social egalitarianism, signalled most prominently by the use of the informal *tu* form of address amongst comrades. This was a striking break for those coming from a rural environment, where non-reciprocal speech codes (using *voi* and *lei*) were obligatory to all strangers and townspeople. There was also discipline and hierarchy in the PCI, but not a great deal of deference, while the rank-and-file on the whole respected their leaders, most of whom in this period came from the resistance and from the same background as themselves.

There were of course degrees of commitment in this social world. At a minimum, party members attended meetings and all bought the newspaper – *L'Unità* – which functioned as a public marker of belonging. They also read it – despite its notoriously esoteric linguistic style – and found that, like all newspapers, it was premised on its readers constituting an 'imagined community'. This was not Anderson's community of the nation, but a community of workers; and it was on that basis that national and international events became news. At a higher level of commitment were party activists who had attended residential courses at one of the PCI schools, and had the crucial role of mediating local and national politics. Men dominated the leadership ranks, and these were expected to devote most evenings and weekends to organisational activities. Members valued leaders who were *preparato* (imbued with knowledge about party history and strategy) and *coerente* (living by the precepts that they advocated). The acclaimed personal qualities of the leadership – their austerity in rejecting a 'bourgeois' lifestyle and the dedication of their lives to the party – were essential components of this moral order. The party was something one could dedicate one's life to; it was a cause – whether one was a leader or a follower – and in this way a political identity was located in a collective and personal historical narrative.

Identity as a comrade was forged in the practices of solidarity, unity, loyalty and discipline in the rich life of the party section, the *Casa del Popolo*, the *feste*. This political culture was created through a social pattern which brought a spectrum of people together outside the workplace, and its decline was due in part to the spread of television and the domestication of leisure and consumption. Until its demise, the communist identity was generated in a narrative of who 'we' were, and who 'we' were opposed to, located in a growing history of intervention in Italian political life. This narrative was still locked into the metaphor of the road which we first encountered in Sesto San Giovanni; but the emphasis had shifted, and as the road had lengthened it had acquired a past as well as a future. There were still subtle and pervasive metaphors constructing political divides in terms of progress: the workers and their party were nearest to the future, the most advanced sector of society. For example, as the great rural exodus filled industrial cities with southern migrants, they were said to be 'workers like us', but from a more backward political culture of deference and deception. But the communists' own identity narrative became weighted with tradition and ritual. Members would learn about the history of their local party, and stories about their national leaders into which were compressed complex political meanings: Gramsci in prison, Amendola organising the northern resistance under the noses of the fascists, Togliatti at Salerno. Leaders elaborated a party line on the basis of insights from their predecessors, and their own authority on the basis of intellectual pedigree. As a demonstration rolled through the streets of any city in the 1970s, the rank-and-file would roar, 'We are communists, the party of Gramsci, Togliatti, Longo, Berlinguer.'

Inclusion and exclusion work together, and communist identity was strengthened and stabilised by a second factor. The virulence of the Cold War created an international fault line between eastern and western Europe – and no less so within societies of both east and west, in each case redefining national identity and the spectrum of legitimate political aspirations. In Italy the dominant nationalist discourse was defined around the country's location within the free west, and for the first time around its Catholic culture. Those who were opposed to this western and Catholic trajectory were represented as subservient to the nation's enemies. This had several consequences for the class politics we have been following. Firstly, the contest became one between Catholic and communist political forces, and was often articulated at the level of ideology in terms of freedom, religion and morality. Secondly, the outcome of all class movements hinges on a struggle over key institutions at the national and state level, but during the Cold War these national institutions (the agencies of the state, including the army, then secret services, and the Catholic Church) were themselves locked into, and responsive to, an international order which included the NATO alliance and, above all, the United States. The result was a substantial widening of the political terrain, so that the outcome of a local struggle by share-croppers to achieve a comprehensive and equitable land reform was dependent on those larger factors which determined whether the PCI would take power: the capacity of the church to convince voters that it was impossible to be a Catholic and a communist, and the resolve of the United States to intervene in its sphere of influence. It was of course the role of the mediating *cadres* of the PCI to explain those higher-level factors, and to translate local demands into a strategy appropriate to the national and international context.

It is important to stress both that these national and international factors fed back into local politics, and that the polarisation of Italian society and broadening of the political agenda were played out on the streets of small Tuscan towns. The aesthetics of a building, the celebration of a wedding, the value attached to local identity – all might become the focus for competing claims about the truth of different political ideologies. The arguments which erupted in public places enacted a kind of political schismogenesis: they presupposed a world divided into two camps, contrasted as positive and negative, with the middle terms and areas of shared experience excluded by the dynamic of the argument. The most banal remark could trigger an argument whose content focused on the errors of the opposing parties: a comment on the deficit of the local administration led to a reply about the obscurity of Vatican finances. Such discussions could escalate across the whole range of the opponent's political world: 'In Russia you are not allowed to criticise anybody's finances', might be countered with, 'If you looked at what the church did during the Inquisition, you would not lecture anybody on liberty.'

In these kinds of exchange the speaker was held responsible for all the actions within their chosen political world. For a Christian Democrat this included the Catholic Church and the Pope, conceived of as integral parts of Italian society. Communism was represented as a unitary phenomenon, realised in the Soviet bloc to which local communists were loyal and obedient – so that Russia, an external power, was the main point of attack, since it revealed the true nature of communism. These exchanges rarely followed a linear theme, but had built into them a tendency to escalation, since each speaker would attempt to move to the highest possible moral ground, reflecting the dominant discourses of the Cold War, which portrayed these political differences as contests between good and evil. They involved particular political skills – the ability to move sideways when boxed in, to find unexpected ripostes, and to have the last word, thus exiting from the escalation. Success required a deep knowledge of political history: if, in the middle of a referendum campaign, your opponent suddenly stated that Togliatti had been opposed to legalising divorce, you could lose a great deal of authority by showing surprise or by not having an answer. There was theatre in these encounters: Tuscans value verbal skills, *finezza*, and such exchanges may have owed something to traditional forms of verbal duelling. But there was a great deal at stake when a person, identified with their party, publicly challenged or defended the hegemonic representations of the social order. We might better think of this as a form of Geertz's 'Deep Play'.

CONCLUSION

As a political subject, the PCI created a complex 'communist identity' which, Li Causi argues (1993: 96ff.), was built precisely on the diverse social positions of its members (rural and urban, manual labourer and professional, employee and self-employed) and the establishment of solidarity on the basis of common values. Living this communist identity did not obliterate membership of any other social group or category, but was higher than any of them: loyalty to the party conditioned solidarity with a share-cropper, or a fellow-townsperson (*compaesano*) who supported a rival party. Togliatti's mass party of the post-war period created a political identity of 'we comrades' (*noi compagni*) that was more complex than the proletarian identity of Sesto San Giovanni. Two conclusions can be drawn from this – the first specific, the second more general. Firstly, the PCI did not create a socialist Tuscany, but it was the movement which produced an historic transformation of peasants into citizens, able for the first time to claim rights and participate in a political and civic culture. This was not produced by the unrolling of some inevitable process of modernisation. It was fought for – and those familiar with the period will know that it was strongly resented and resisted by – entrenched urban power, and the paternalistic condescension of Catholic integralism, with its vision of rural segregation. As a process it should be assessed on the same terms as other civil rights movements.

Secondly, even a compressed summary of this period illustrates the variety of processes which shape the direction taken by a political movement. It is not that the lines of class division were particularly complex; they were in fact clearer here than in any of the other examples used in this book, etched in the landscape. But those social divides did not map cleanly onto the central political fault line in post-war Italy – that between the Catholic and communist movements – since both sought in various ways to establish property rights for share-croppers. Their struggle for autonomy was incorporated into a wider Italian political battle, itself heavily conditioned by the dynamics of superpower rivalry, with decisive consequences for themselves and for the political culture of the region.

SOURCES

There is a very large literature in English on share-cropping and the rural economy in central Italy. Silverman's work in Umbria is based on research while the *mezzadria* was still extant: her 1975 volume, *Three Bells of Civilization*, was a pioneering work in terms of its combination of

anthropological and historical data, and there are a number of other
articles (Silverman 1965; 1970; 1977). Gill (1983), Lyttleton (1979)
and Snowden (1972; 1979) deal with the first stages of political mobilisation and the fascist reaction. I have also drawn on my own research
in southern Tuscany on economic and political change (Pratt 1980;
1987; 1994). There is an even larger literature in Italian, including
Sereni (1947) who began the long debate about the *mezzadria* and the
transition to capitalism, continued in Giorgetti (1974), with more recent
contributions in Pazzagli *et al.* (1986). I am particularly indebted to the
anthropologist Pietro Clemente, whose perspective on rural society (some
published: Clemente 1980; 1987) challenged my own thinking.

There was enormous interest in the Italian Communist Party in the
1970s and 1980s on the part of historians and political scientists,
including Sassoon (1981), Davidson (1982) and Boggs (1986). A second
industry grew up around the work of Gramsci: Lumley (1977), Hall
(1996) and Urbinati (1998) give some idea of the variety, and provide
further sources. The complex relationship between 'official' communist
narratives and rank-and-file interpretations of history is explored in a
variety of sources, from Portelli (1990) to Periccioli (2001). There are
also anthropological studies. I have drawn on my own research on
political cultures, some of it published: Pratt (1989 and 2001) deal with
political language and practice, while Pratt (1986) includes a more
detailed analysis of Cold War polarisation. Kertzer (1980) is a very
accessible account of grass-roots social and political life in Bologna; his
1996 study is based on a wider range of sources and adopts a more
critical perspective – both will be enjoyed most by those who are
primarily interested in politics as symbolic action. Shore (1990) uses a
multi-disciplinary approach, including his own fieldwork in Perugia, to
trace the history of the PCI from 1921 until its dissolution. This is a very
well documented account, which takes communism to be inseparable
from democratic centralism, a viral infection first acquired from the
Bolsheviks. These studies have many strengths, but do not convey much
of what party members and supporters gained, or lost, from their mobilisation in this cause, beyond the company of others. Li Causi (1993),
concentrating on a small area (Siena province) gives the reader some
glimpses of the loyalty and passion, and a sense of a society in movement.

5   A SHORT HISTORY OF THE FUTURE

The previous three chapters on Sesto San Giovanni, Andalusia and Tuscany examined revolutionary class movements, and before we move to the politics of ethnicity and nationalism, this short chapter will take stock of what has emerged so far. It will deal in turn with the processes of mobilisation and the issue of identity, in order to suggest some common themes which run through class movements, as well as comparisons between them.

First, we have to return to the term 'class'. There are debates about the theorisation of class as an economic category (defined either as a layer within a stratified structure, or as a relationship within a productive system); there are debates about classes as collective political actors; and above all there are debates about the relationship between economic categories and political movements. Within the Marxist tradition they centre on the famous distinction between class-in-itself and class-for-itself. One of the most important discussions emerged out of E. P. Thompson's analysis of *The Making of the English Working Class* (Anderson 1980; Kaye and McClelland 1990; Thompson 1963; Wood 1995). One approach sees the political as a direct reflection of the economic: the capitalist economic system creates classes – people who occupy the same position in a production system; they then do either become conscious, or do not, of that position, and articulate their shared interests. When classes acquire 'consciousness' they become conscious of their economic position, the interests they articulate are essentially economic, and the composition of the resulting political movement is congruent with the economic category of its members, give or take a few intellectuals who provide consciousness and direction. Class as a political term is, in this view, coterminous with its economic counterpart. It is easy to use shorthand phrases (such as 'the Italian Communist Party mobilised the working class') which reproduce this formula.

Others object to a framework which takes class as a given and prefer to see class as a more purely political phenomenon. In breaking with the reductionism of some Marxist models and formulations, some writers then go on to minimise any connection between the way productive systems structure social relations and political movements – instead, class is constructed discursively. 'The ultimate conclusion of this argument must be that a caveman is as likely to become a socialist as a proletarian – provided only that he comes within hailing distance of the appropriate discourse' (Wood 1986: 61; see also Gledhill 1994: 186).

These are polarised positions within a debate about class, and it is easy to draw the conclusion that neither economic reductionism nor (post-Marxist) idealism will reveal the dynamics of class movements. We need an approach which explores the interaction of complex economic and political processes, and though such an approach does not produce simple formulae, there are some general points which emerge from the preceding case studies. Firstly, capitalism creates a variety of production systems, and many forms of socio-economic inequality: those based on wage-labour are historically very important, but we should not assume that 'the working class' is synonymous with the proletariat. The working population includes people who gain their livelihoods from a variety of activities, and whose lives are shaped by a complex mixture of autonomy and dependence. Secondly, in order to understand the strategies for gaining a livelihood outside (or alongside) wage–labour relations, we need to think critically about what we mean by 'the economy' (and 'economic interests'). In contemporary society 'the economy' tends to be a category which covers only monetised relationships, a sphere regulated by its own autonomous laws or rationality. Narotzky (1997: 220) has pointed out that it has come to refer to only a small part of total social reproduction, but has nevertheless become the general framework for the interpretation of social life.

Thirdly, we have seen that class boundaries – the lines of opposition – are a response to the way production is organised, but also, and more specifically, to the way each movement interprets the configuration of interests amongst the different actors within the economy. Are the artisans, the share-croppers, the destitute, with us or against us? However, the political construction of class is most dramatically revealed in the repertoire of action each movement adopts – strikes, occupations, armed insurrection – and the fault-lines which follow from them. Finally, the protest embodied in class movements is not confined to issues such as wage levels and living standards. Each of the movements articulates around a much wider critique of the existing social order, and contains a wider vision of what it is to be human and of how society might be organised to create greater justice, equality or autonomy. In doing so they have to address other kinds of identity – those involving gender, territory or religion for example – and incorporate them within the master narrative of class. Out of these wide-ranging critiques and visions, such movements generate political cultures, in part through addressing and reacting to rival political cultures: think of the impact of anarchism and Christian Democracy on Italian communism. I will elaborate on these points as we go through the case material.

I have dealt only with movements which conducted a sustained attack on existing property rights and the state, but even within this more limited range 'the working class' emerged as quite a variegated entity. There was the struggle of the urban industrial proletariat, but also widespread mobilisation in rural areas, amongst seasonal wage-labourers and share-croppers. In southern European regions the majority of the population lived and worked in rural areas until late in the twentieth century (see Table 9.1 on p. 188). Many of those regions had very radicalised and anti-clerical populations, and were major strongholds of revolutionary politics, contradicting stereotypes of rural areas as traditional and conservative. Each movement had at its core one group which formed the majority of the population and whose demands dominated local politics, but they were never alone on the scene. Ethnography reveals, for example, the importance of economic operators who, though technically self-employed, generated their income in the ambit of the major production centres: marketing, supplying, servicing. There were households of peasants, artisans and shopkeepers who gained part of their livelihood from seasonal wage-labour or as outworkers. These economic operators and mixed-income households are not well captured by employment statistics, with the cat-egorisations of work they use (Smith 1991). The same problem applies even more strikingly in the case of women, who often disappear from official labour-force statistics: farm women, for example, who become 'housewives' and 'inactive' in the censuses. Whole theories about stages of capitalism and the emergence of post-Fordist and post-industrial society turn on the relative presence in the economy of groups who have mixed-incomes, or are 'self-employed', but sometimes on the basis of very unreliable evidence.

At the most general level the radicalisation of these populations can be attributed to the development of capitalism: changing property rights; concentration of the ownership of resources; enlarging markets; the destruction of older forms of livelihood and their associated skills. At the same time, we have to acknowledge both that other factors shaped this radicalisation, and that people lost control over the conditions that enabled them to gain a livelihood in a variety of ways. The result was not a simple economic category but a variety of economic figures, amongst whom the classic industrial proletariat were a minority. In studies of politics in the 'developing world' it has often been pointed out that models of class struggle based on European industrial models are inappropriate, and the same argument can be extended to much of Europe itself. The class movements described were alliances, in at least two ways. Firstly, they built local alliances amongst this variety of economic figures, sometimes widened to include the self-employed or the unemployed and destitute, sometimes excluding them. They were built in different ways according to social conditions and the political discourse which interpreted those conditions. Politics itself created processes of inclusion and exclusion and shaped the lines of class division. One of those discourses was Marxism, and we shall return to this point. Secondly, the movements built forms of solidarity and identification, translated into co-ordinated action, between workers in different locations. No economic system 'assembles' people at any level above that of the production unit: the factory, mine or estate. Only politics can do that.

Mobilisation took place in the workplace and outside it: unions fought over pay and conditions, and sometimes made more radical demands over control in the workplace. This level of action was itself difficult to organise, since employers did not recognise unions and allocated neither time nor space for meetings. Unionism itself, moreover – whether based on trade or sector – reproduced some of the lines of division of the organ- isation of production. It should also be stressed once again that there is no guarantee that sharing the same economic position – as factory hands, estate workers, share-croppers – will generate either solidarity or a political movement. On the contrary, there were normally consider- able areas of competition and lines of division within these categories: between skill grades in a factory, between *pueblos*, or between estates. It was the work of the political movement, operating in a fractured landscape, to subordinate – or attempt to subordinate – these divisions into a larger political project.

Workplace mobilisation was vulnerable and limited unless linked to wider political action outside work time and outside the workplace. Movements took root if activists created and maintained organisational structures appropriate to the specific economic and social environment in which they operated. In Sesto San Giovanni, these were the social and political *circoli*, and later the party section. In Andalusia, where there were no stable employment contracts, and where there was a complex relationship between the *pueblo* and the estates, it was the *sindicato* or *centro* which organised both workplace and *pueblo* politics. In Tuscany, where there was most atomisation and no specific workplace, party activists built on existing patterns of socialisation, choosing strategically located farmhouses where evening meetings of a circle of share-cropping families could be held. As mobility increased, the Casa del Popolo – built and run by party members – became the political and social centre in the small provincial towns. They were remarkably successful both as channels of communication and debate, and in creating the habits of solidarity amongst people from different social backgrounds. All these movements, from the workers' libraries of Sesto to the anarchist newspapers of Andalusia, generated an upsurge in literacy and in publication.

This is the vital infrastructure: meeting places under the movement's control, both furnished and resourced; newspapers, and later a whole battery of other ways of communicating – the cyclostile machine, photocopiers, loudspeakers – employed to generate an autonomous source of information which was very explicitly an alternative to that of the 'bosses' or the state. It was this infrastructure which enabled a Breda foundryman to meet those in the paint shop; or in the 1970s allowed the PCI to pull a million protestors onto the streets within 24 hours of a bomb outrage. Equally important was the way in which alliances were constructed between the dominant category of workers and other categories (such as artisans or the unemployed) who were present in the locality. The extent to which this happened, and how it happened, depended on the particular class discourse which articulated the movement, and on the way in which lines of cleavage (between skilled and unskilled workers, between share-croppers and labourers) were reinforced or mediated. All the movements involved some kind of alliance, and some form of general strike was employed by all these movements at critical junctures, whatever their political labels. Shutting down all economic activity in a town – factories, workshops and shops – occupying the streets and challenging the capacity of the state to control public spaces were all ways of constituting a movement through action, and creating alliances on a territorial basis.

The leadership of these movements was dominated by men. Women made up about one-third of the labour force in both the factories of Sesto and the Andalusian estates, and were an integral part of the workforce on Tuscan farms. They participated in workplace struggles and public demonstrations, but they very rarely broke through into leadership roles within the movements. The reasons for this are complex, and include the ways in which waged employment and the identities which derive from it are constructed in gendered ways, and mesh with domestic roles. We should also note that the key political forum was not the workplace but the endless round of evening meetings in the sections and *circoli*: the higher the responsibility, the wider and more time-consuming became the political round. This level of activism was normally incompatible with domestic obligations, which were in turn a consequence of the wider gender order.

In all cases political mobilisation operated in at least two contexts: the workplace and the local territory. It is worth underlining the fact that all these class movements developed organisational forms based on territory – what have been called communities of resistance – and also attached a great deal of importance to the control of local public spaces. Anarchism stressed territory more than the other movements: it was the rights of a localised population to liberty and economic equality which were the purpose of the political movement. This is evident in the dual meaning of the '*pueblo*', which generated a particular conception of class and place. Beyond the social space of a town or district, class movements operated at the national level: their strategies were determined by the conflicts operating throughout a national territory, and by the strategic goal of seizing or replacing the machinery of the central state. Again it is anarchism which is unusual, in that it developed loose federations to co-ordinate general strikes, but aimed to destroy the nation-state rather than constitute its government. There are important questions about the connection between local and national objectives, and we can explore this further in looking at the development of class discourse.

Living and working in polarised social environments characterised by hardship and subordination did not inevitably generate 'class con-sciousness'. 'If classes are to appear in politics they must be organised as political actors. Again, political class struggle is a struggle about class before it is a struggle among classes' (Przeworski 1977: 372, quoted in Foweraker 1989: 258). In other words the first political development is the establishment of some version of a conceptual framework of class in an environment where there were always other ways of interpreting experience. One alternative was an ethnic interpretation of social position, the crucial element which Gellner believed explained the presence of sustained revolutionary movement, though strangely it is absent in the three cases presented. However, there were other discourses circulating in these social environments. Oppression or direct expropriation of surplus could be naturalised as part of a hierarchical order, and in that case class discourse was a rupture with organic models of society, as we saw with the *mezzadria*. Alternatively they might be naturalised as part of an inevitable disorder – society was a competitive jungle such that social position was experienced as the direct result of personal achievement or failure. Even in the midst of class movements there were always some who followed other ways of affirming or denying human agency.

In the movements examined, class discourses contested the organic and competitive interpretations of society and provided an alternative view of the existence and history of social divisions, and sometimes of their connection with human nature; that is, they offered an understanding of class belonging (of who 'we' are), and how society came to be structured according to class. In doing so, they also articulated a radical reversal of the normal, everyday understanding and experience of the relationship between the poor and the rich, the labourer and the employer, suggesting that the lines of dependence in fact ran in the opposite direction, and that politics could turn the world upside down. The moral charge of a general strike derives from it being an act of revelation: it shows the world as it really is, and prefigures how it will become after a revolution.

Both popular and more abstract versions of these discourses existed, but rather than cover the whole field, I have concentrated on the part which constituted identity narratives, suggesting that these identities are best understood as complex, politically constructed narratives, defining boundaries and oppositions, positioning collectivities in social processes, as well as in time and space. The most familiar, and the most narrowly defined, of these narratives was that which emerged amongst the early industrial proletariat of northern Italy, and was heavily influenced by Marxism. It interpreted capitalism as built increasingly on a mass of workers selling their labour – people who had no 'past', no skills and no property, whose identity was defined around who they would become after the revolution. They represented the future, while the ranks of skilled labourers, who had a past in the tradition of artisan work practices and mutualism, were counted as 'property-holders' and considered susceptible to 'reformist' economic demands and liberal democracy. This 'purist' line on the revolutionary commitment of different categories of workers was contested within the Italian Communist Party, and was not, in the long run, Gramsci's own position (see Hall 1996). In political practice alliances were forged; however, there was never any doubt that the proletariat was central, and that diversity was interpreted in terms of a Janus-faced narrative positioned around the revolution: the proletariat was the future; all other groups were becoming more like the proletariat; after the revolution there would only be a proletariat.

Other movements may have had a conception of 'those who work' at their core, but they were not necessarily built around a theory of wage-labour and the historic role of the proletariat. This emerged in the discussions of the connections between work, the person and identity found in the anarchist movement in Andalusia. Here, as in other rural regions, the opposition was between those who did manual labour and those who lived off the income of their properties, sometimes exhibiting an ostentatious disdain for getting their hands dirty. So the anarchist movement was built around the field-workers, but operated mostly in terms of a localised and inclusive dichotomy between the parasitic rich and the working poor. It pitched labourers against landlords – both in relation to employment, wages and conditions, and because it was the landlords' property rights which stood between the labourers and a just society. The struggle was widened through the *pueblo* sections to include all those who did manual labour or aspired to an egalitarian society, and extended to an attack on all those who perpetuated authoritarian relations and stood around with their hands idle, including the police and the priesthood. It was the general strike which created the faultlines of class relations, drawing in and polarising all the middle strata (artisans, peasants, muleteers) into the struggle over land.

What these comments suggest is that, politically, class is constituted by the movement and its discourse. The movements interpret and build on the divisions and configurations created by the economic processes of a capitalist society; but once constituted they are also shaped by the dynamics of a purely political process. This is true both of ruling coalitions, and of the class movements which oppose them. We can elaborate on this observation through the third case study, that of Tuscany.

After 1945 the PCI maintained many features of the orthodox communist tradition in its practice of democratic centralism and its commitment to a revolutionary process which would achieve 'the dictatorship of the proletariat'. But it was also a mass party, which mobilised a far wider spectrum of the population than the proletariat, and sought to achieve an electoral majority. The historic bloc which the PCI attempted to create was defined, primarily in economic terms, as all those opposed to 'monopoly capitalism', but as a political movement it operated on a terrain where the central issues and lines of cleavage were defined rather differently. Italy's strategic location in the Cold War, and the PCI's real and perceived subservience to the USSR, both 'internationalised' domestic politics. On one side monopoly capitalism was partly identified with American imperialism, while on the other those seeking its overthrow could be portrayed as aliens threatening the Italian nation and its cultural traditions. The class movement rapidly triggered competing discourses about the Italian nation – discourses which were themselves being reshaped by the Cold War. However, the most dramatic shift in the political terrain was generated by the Catholic Church. The PCI tried intermittently for 30 years to create alliances with rural producers and middle strata who were politically organised by the church, but it encountered intransigence and failure. It was put firmly on the defensive by a Catholic movement which represented the PCI as the enemy of liberty, the family, and all Christian values. The class movement was reshaped around the competing national and moral discourses which constituted Catholic and communist worlds within Italian society.

The international context is domesticated into the movement at many different levels: it affects the strategic choices and resources available to a national leadership, but it also affects everyday politics in the workplace and the streets. After all, actions like the Pope's anathema against communists, and the policy of exclusion which derived from it, were only effective if they operated in the local context of villages and the urban districts, and the example of Cold War rhetoric in the last chapter was designed to show this in operation. This brings us to the final point. Class divides are embedded in sets of social and cultural differences, and in an economic system which includes many 'intermediate' positions, occupied by operators who might be drawn to either side. In saying that class is constructed by a movement and its discourse, we refer precisely to processes of interpretation, inclusion and exclusion, and also to the absorption and modification of those social and cultural divides by the movement itself. In Tuscany the class movement developed in a society which contained long-standing and sharp divides between the urban and the rural, the devout and the anti-clerical. It incorporated existing cultural discourses and oppositions: between the civilised and the natural, autonomy and dependence, individualism and solidarity. Within a class movement we find interactions between organisational levels and complex transformations of existing social and cultural divisions.

We have seen how each of these movements relied on a territorial infrastructure which was essential for generating activity and solidarity between categories of people who were not 'assembled' by relations in the workplace, including all those who were not full-time wage-labourers. It was also clear that classic forms of action like demonstrations and some versions of the general strike had as one of their ambitions the control of public space, challenging the normality or legitimacy of governmental uses of force. This public space was also the arena for contesting everyday forms of power as manifested in rules of consumption and deference. Indeed, in the rural environments the attack on these customary forms of subservience – 'creating men who cast shadows' – was considered an essential first step before political mobilisation around property rights was possible. Here, in the literacy campaigns, and in the extensive critiques of Catholic teaching and practise (including those on gender in the case of anarchism), we find a much broader agenda than is commonly associated with class politics. It includes challenges to constituted authority in many fields of knowledge, and alternative understandings of power, morality and the person. It is true that all these dimensions of power were formulated in terms of class relations, and their solution seen as dependent on the abolition of existing property rights and the state that maintained them: this is what is meant by saying that class constituted a 'master narrative'. It is also true that these movements were fundamentally involved in cultural struggles (although obviously not the whole range of cultural struggles), and if we cannot now see this it is because we have been blinkered by our own definitions of culture.

So far I have covered some of the issues involved in class construction: the interpretation of economic divisions, the effects of political action, and the wider political and cultural contexts. We can now turn briefly to the issues of identity and morality. The term 'identity' is one of the most problematic in anthropology. Attacking it through the concept of a narrative opens up the space to analyse similarities between class and other kinds of political movement, though there are many ways in which such an analysis can be developed. Passerini's work (1987; 1996), for example, analysing a different kind of material – life-histories – opens up a very complex set of issues, and reveals both ruptures and continuities in the construction of personal identities, a point taken up again in the Conclusion. Most of the discussion in the previous chapters has concentrated on collective identities, partly because the construction of solidarity – 'we' – is so salient in class movements. This bounded collectivity is constructed in the workplace, in the town, and as an imagined community at national and international levels; it is an historical construct reproduced over time through political cultures and institutions. It is also in many ways a highly moralised identity: the boundaries are moral boundaries, while within the anarchist or working-class 'cause' there is an extensive ethical vocabulary of commitment, consistency and betrayal. This is the political identity I first encountered in a Tuscan town in 1970, where every step of the way one could hear, 'Is he a comrade?' (the answer – *Si, e un compagno* – could have half a dozen different intonations).

The stress on collective narratives is also a reflection of the fact that, in this political culture, individuals identified themselves with a collectivity: they found themselves by recognising themselves in others. You very rarely found the first person singular in a speech by a Communist Party politician. If a person broke with the movement, or the movement itself collapsed, this represented a loss of identity. This is echoed in Foweraker's comments on the old guard communists during the post-Franco transition to democracy in Spain, 'who struggled for liberty only to lose their identity in the new freedom and find themselves without political meaning for the first time in their lives' (Foweraker 1989: 169). If a movement collapses, or repudiates its previous objectives, then its adherents lose the narrative within which they have previously interpreted and valorised their lives. A commentator on the decline of the Italian left in the 1980s summarises the situation using terms which are more commonly found in the context of ethnicity: 'It was as if in these days an entire class had lost its own linguistic codes, its own cultural traditions, its own political referents' (De Luna 1994: 21).

The account of politics in Sesto San Giovanni suggested that the translation of economic processes into a moral drama is essential in generating revolutionary politics. This moral reading of existing economic relations is found in all three movements, and goes beyond a view that the state might intervene to ameliorate conditions; instead, the basis of the whole existing system is seen as intrinsically exploitative, alienating, unjust or unfree. However, at this point we also need to make distinctions. Marxism developed a theory of revolution and socialism which claimed scientific status, grounded in rationality and the uncovering of laws which demonstrated the historical role of the proletariat (Przeworksi 1977: 345). It stressed this intellectual heritage as a way of distancing itself from the political culture and practice of anarchism, which was based on utopian visions and moral crusades against injustice. Anarchism operated with a more generalised concept of the poor and oppressed, partly because it emerged in environments less dominated by full-time wage-labour. This was reflected in its political practice, often portrayed as spontaneous and undisciplined – a politics of what ought to happen rather than of what is happening. We can present the distinctions between anarchism and 'scientific socialism' as interconnected and comprehensive: in their claims to truth, in their definitions of class and political strategy, and in their narratives of past, present and future. These differences are evident in the ethnographic and historical material we have examined. There is a connection, for example, between the critique of the present and the issue of progress in the revolutionary narrative – something which emerged in discussing the 'millennial' temper of anarchism. But the differences also need to be qualified, since, in the everyday life of the movements, in Italy or Spain, the contrast is more blurred, with some convergence of people and objectives. What they share is the view that a just society lies in the future, not in the past, and can only be achieved by revolution. This sets them apart from both nationalist movements and most contemporary European politics. We can now turn to the death of class.

If class, in the sense employed here, is constructed politically, it can also disappear. This does not imply that there has been an overall decline in the number of wage-labourers (they have grown in the service sector), or that the concentration of capital and power has diminished (it has probably increased). What declined in the second half of the twentieth century were European revolutionary class movements. Most of the political parties which had articulated revolutionary ambitions had become reformist, working for policies of redistribution and increased welfare provision while incorporated into the institutional practice of the state and the management of the economy. They were often very successful in their own terms – though frustration with reformism was one of the factors which triggered the 'new social movements' after 1968. Lumley (1990) has given us an exceptionally rich analysis of the revolutionary programmes and mobilisation of factory workers and students in Milan in 1969, but that moment passed.

The reasons for this overall decline constitute a problem which falls outside the scope of this book, although some of them can be mentioned. Politically, one of the most important is the most simple – revolutionary movements were defeated. From Greece to Spain, Hungary to Italy, the repressive force of the existing states was able to crush them, usually creating authoritarian regimes and abolishing democracy in the process. Mazower (1998: 3) has noted that in the 1930s democracy survived only on Europe's northern fringes. Secondly, where democracy survived or re-emerged it presented a serious dilemma to revolutionary movements. Strategies which included the pursuit of electoral majorities led to a widening of alliances, and a dilution of a revolutionary strategy with the proletariat at its core, while the fate of the parliamentary party itself came to eclipse the movement which it represented. After the 1950s, the widening gap between living standards in eastern and western Europe, and growing awareness of the evils of Soviet communism, led to the discrediting of the USSR as a model for western class movements.

Economic changes affected the social composition of European societies, creating what some have termed a post-industrial or post-Fordist society, though interpretations of this process have generated very sharp disagreements. Decentralisation or subcontracting in the production process was scarcely new: it had been normal for a long time in agriculture, construction and the textile industry (Narotzky 1997: 215). What was important was the decline or relocation of mass production in certain key sectors. As the Andalusian or Tuscan estates empty, mines and shipyards close, and the big assembly plants are mechanised, there is a dispersal of the critical mass of workers around whom a class movement is aggregated. The construction of a class movement in those locales depended on a high level of activism outside the workplace, in a dense infrastructure of meeting-places and communication networks. One of the most striking social phenomena since the 1960s has been the privatisation of leisure time, and especially the comprehensive diffusion of television. This not only dealt a serious blow to traditional forms of activism, it propagated a very different political culture. With a few minor exceptions, the movements which had leapt on the opportunities offered by print singularly failed to gain a foothold in the new media.

It has been argued that there has been an accommodation to the way production is organised. Workplace struggles concentrate on wage-levels – the distribution of the surplus, instead of control over production. This represents a trade-off, and with rising living standards energy switches to achievement and identity construction in the field of consumption rather than production. The interesting thing about this argument is that it has been applied to much earlier periods than we have been considering here. Bauman (1982) develops it in his analysis of nineteenth-century England, and without entering into complex historical generalisations we can point to two themes in Bauman's work which are relevant to the south European examples. Firstly, political rad-icalisation is very often produced by loss of autonomy in the sphere of work, loss of control over the labour process and the product. Accom-modation and the incorporation of the labour organisation into the capitalist system 'were accomplished through the economisation of the conflict ... [that is] the substitution of the wage-and-hours bargaining for the initial conflict over control of the process of production and of the body and soul of the producers' (Bauman 1982: 100). Secondly, the narrative of working-class identity is not monolithically about the future; it has within it, to a greater or lesser extent, precisely a memory of loss, and the desire to re-create a past condition of autonomy. We shall return to this point in the Conclusion.

The end of class movements signifies the disappearance of that process of co-ordinating action between factories, localities and occupational groupings in order to achieve a society-wide socialist transformation. Movements which began as class movements can of course survive even if their class discourses atrophy, and they transmute into something else. Sewell suggests that 'While class discourse may commonly have been more durable than class institutions in the nineteenth century, class movements have sometimes outlasted class discourse in the twentieth' (Sewell 1990: 73). In post-war Tuscany reforming and revolutionary strands co-existed, but the overall construction of class identity around 'who we will become' slowly gave way to an identity constructed around a lengthening cultural tradition. In the celebration of the history of the party and its leaders, the sense of class as a relationship – as oppositional

– became less evident. Local movements in the workplace or in a union can also survive as forms of resistance, but when not harnessed into a transformative political project they may become an inturned corporatism – an uncoordinated struggle over wage-levels by isolated groups of employees. This kind of action has fed the assertion that class politics is about economics (in the restricted sense that Bauman identifies), since, when a general and collective programme disappears, so too do all those mediations between local and national levels, and the complex transformations of existing social and cultural divisions. In the next chapter we will turn to nationalist movements.

6   THE BASQUE COUNTRY: MAKING PATRIOTS

In accounts of nationalist movements the analysis often seems to be at war with itself, since the categories used in the writing appear to assume the historical continuity and stability of nations, while the argument itself is concerned to deny this. For that reason it is worth stating at the beginning of this chapter the main premise which informs this argument, and the argument of many (though not all) of the sources from which it is drawn. There was no stable, bounded or homogeneous ethnic group called the Basques who existed, awaiting nationalist leaders to mobilise them in opposition to another stable, bounded and homogeneous nation called Spain. Instead, when at the end of the nineteenth century a Basque nationalist identity began to be articulated, the population of the Basque provinces demonstrated a high degree of heterogeneity with a series of cross-cutting economic, political and cultural cleavages. There were no neat dichotomies, but a system of differences. What the Basque nationalists did was to select out certain differences which served as 'ethnic markers' (and the selection varied over time), and weld them together. In welding them together they created something new, and it was new even if some of the markers and cultural differences out of which it was constructed had existed for a long time within local society. What was new, viewed from the inside, was the affirmation of an ethnic group; viewed from the outside, it was a set of ethnicised social relations. These relations consolidated, evolved and transformed the face of political life both within the Basque territories, and between those territories and the rest of Spain. Nationalism presents itself boldly with nouns and essences: we need an analytical vocabulary of verbs and relations. The question driving this chapter is not 'Who are these people, and what sets them apart?', but that of the political process whereby certain economic and cultural differences, some of great antiquity, were constructed as ethnic.

This chapter is organised into two parts. The first and longest part deals with 'ethnogenesis', the emergence of nationalist discourse and the consolidation of ethnic boundaries in social and political life. The complexity of Basque history and the richness of the ethnography make this an ideal case study for the analysis of ethnicity in both its dramatic and its everyday manifestations, and for the exploration of ways in which the significance of cultural differences is embedded in economic and social relations. This part will start with an account of development in the Basque provinces, up to and including the rapid industrialisation of Vizcaya at the end of the nineteenth century. This leads into an examination of the emergence of nationalism at the turn of the century amongst the urban middle classes, looking at both identity discourses and the range of activities and organisations which constituted the nationalist movement itself. Attention then switches to rural society. At the beginning of the century Basque nationalism, although built around a rural ideology, was seen by most rural people as an urban invention, at odds with their view of their own trajectory; 50 years later it had entered and ordered their conceptions and practices.

The violent repression suffered by the Basque people during the Civil War and its aftermath led eventually to some significant innovations in the political directions and strategy of the nationalist movement, and specifically to the emergence of ETA. The second part of this chapter deals with the revolutionary ambitions of Basque nationalism, and examines two further themes: the effect of violence on political mobilisation, and the attempt to combine class and ethnic politics.

PART ONE: NATIONALIST DISCOURSE AND ETHNIC BOUNDARIES

From Foralism to Industrialisation

In the sixteenth century, at the height of its glory, Spain's ruler was called King of the Spains: a recognition of the diversity of both the colonies (the Netherlands, the Americas), and of the regions which made up the Iberian peninsular. Spain was plural, and there were many ways of being Spanish. By no means the least of them was to be Basque, since this northern mountainous area had never been incorporated into the Moorish Kingdoms, and formed one of the strongholds from which developed the Christian Reconquest. The acclaimed attributes of the people from this region – 'Strong in arms, pure in blood, steadfast in religion' – made them paragons of Spanish virtue. The provinces claimed by the Basque nationalists in the twentieth century (four in Spain, three in France) at no time constituted a political unit with its own central administration. Partly as a result the language, Euskera, remained uncodified, an oral language of often mutually unintelligible dialects used only by the rural population and by the lower clergy.

If the provinces had never constituted a political entity as such, the people who lived there did maintain or acquire certain distinctive political rights. One of these was the recognition that the people of the provinces had 'collective nobility': because they had never been contaminated by Moorish or Jewish settlement they had pure blood lines, and hence were eligible for office in the church and the state without the need to demonstrate individual pedigree. Secondly there were the *fueros*, 'collections of local laws and customs together with specific economic and political immunities underwritten by the Kings of Castile (and later of Spain) in return for political allegiance to the monarchy' (Heiberg 1989: 20). There were no standardised *fueros* for the Basque provinces; they were written in Spanish, and the existence of local statutes was general throughout Spain (and much of Europe). However, the *fueros* of the Basque country gave more extended rights, created greater local autonomy, and lasted longer, than those of other Spanish regions.

The major line of cleavage within the Basque provinces was between rural and urban society. The classic representation of Basque rural society was of an independent peasantry, occupying a small farmstead (*basseria*), practising impartible inheritance, largely self-sufficient but also reliant on access to considerable areas of church and communal land. But this generalised picture needs substantial qualification: firstly, in that it holds more for the mountainous northern provinces of Vizcaya and Guipazcoa than for the drier inland province of Navarra, where a different agrarian regime operated: secondly, in that there had existed a substantial market in land for many centuries, and a notable oscillation between periods when land was concentrated in a few hands and the majority of the farmers were tenants, and those periods when ownership was more widespread. The farmstead may have generally been indivisible, and a symbol of continuity, but its ownership was not. There were tensions within rural society between landlords and peasants, and by the mid-eighteenth century there emerged a first defence of certain social and cultural principles in the work of a Jesuit called Larramandi, who lauded the purity, nobility, austerity and egalitarianism of hard-working peasants. This was not a description of the actually existing rural society, since in 1750 the powerful and parasitic landlords dominated; it was a eulogy to how things should be.

The non-inheriting sons and daughters of the *basserias* moved into the towns, or down to the fishing and whaling communities on the coast, or emigrated to the Americas. Basque urban settlements (*villas*) were military and commercial centres, organising part of the trade between Castile, Europe and the New World. They contained their own professional middle classes and were also centres for artisan production, including that based on the timber and iron of Vizcaya, though the foral regime prevented its export. The prosperous families of these centres also periodically invested in land, and they conducted all their business in Spanish.

Both sectors of Basque society were transformed by the two Carlist wars (1833–39, 1872–76) which were triggered by a conflict over the succession to the Spanish throne, but involved a complex of social forces within Spanish society. Generally, the forces of liberalism and urban society were in opposition to those favouring a rural, decentralised Spain united around the monarchy and a strong, purified church. Liberalism won, and the settlements of 1878 consolidated the trend of the previous decades. As we saw in the discussion of Andalusia, both communal and church lands had been sold off, hitting both the clergy and the peasantry, a situation aggravated by the abolition of the *fueros* and the establishment of a regime which taxed land and livestock more than industry. Administration and public order became more centralised, while all customs posts were moved to Spain's national frontiers, and a very high protective tariff system put in place.

The result, especially in Bilbao and Vizcaya, was one of the fastest periods of industrial growth seen anywhere in Europe. British capital was invested in the iron mines, both for export and for local industry. This growth created a sprawling industrial landscape of mines, blast furnaces, engineering works and shipyards, with the labouring population lodged in company barracks, suffering low pay, overcrowding and frequent epidemics.

At the end of the nineteenth century the Basque provinces contained a great variety of social circumstances and political aspirations. The rural areas untouched by industrialisation were home to a population of small farmers, mostly Basque-speaking, staunchly Catholic, and struggling to adjust to the loss of communal land rights and the new market conditions. An industrial oligarchy ruled in Bilbao: the bulk of the iron and steel industry, and the flourishing banking sector, were in the hands of half a dozen families. They controlled the local political machine, but their immense power was exerted primarily on the national stage, since they were major players in Madrid's liberal government circles, and able to shape Spain's economic policies. Next there was a rapidly growing urban proletariat, which was internally divided. Those recruited from the rural hinterland tended to cluster in the smaller firms, while those who had migrated from other regions of Spain formed the majority of the workforce in the mines and steelworks, and also dominated the lower ranks of the local administration. These migrant workers created one of the first and most militant socialist movements in Spain. The socialist

PSOE was founded in 1890 in Bilbao, which became the centre of a series of violent general strikes over the next 20 years.

Between the oligarchy and the proletariat existed various middle strata. In the industrial boom after 1876,

smaller firms were absorbed or driven out of business by the large industrial combines. The decline of this sector of Basque leadership was reflected in the shift of their sons from industry into the professions. They were lawyers and doctors, journalists and teachers, artists, composers and writers, the providers of services, such as transportation, communication, design and planning. They were the second generation of the industrial boom. (Clark 1979: 38, quoted in Conversi 1997: 56)

They were the losers in the transformations which had swept through the Basque provinces. Their economic space was eroded by the industrial and financial giants which had emerged in Bilbao, and by the abolition of the foral regime and the status of 'free-trading' zone which the provinces had enjoyed. Their political role was reduced by the oligarchy's local and national control, while their livelihoods were threatened by the possible encroachment of militant socialism into the smaller workshops run with labour from the rural hinterland. Carlism was a reduced force and belonged to the past; the liberal revolution was the cause of their present problems, while socialism was the future threat to their interests. They needed a political project, and they created the Basque nationalist movement.

The First Stages of the Nationalist Movement

Sabino de Arana (1865–1903), the founder of this movement, was typical of this generation and background. He was born into a Catholic family in Bilbao, industrialists whose fortunes had suffered from their identification with the defeated Carlists. The starting point for understanding this movement has to be the sense of loss which men like Arana experienced when contemplating their own fortunes and the kind of society that was emerging so rapidly around them in the 1890s. The dark satanic mills of Bilbao and their associated squalor had eaten up the *basseria*'s rural hinterland, creating a booming city of migrants: only 20 per cent of the population was born in Bilbao, while nearly half came from distant regions of Spain. Euskera as a language had disappeared in the city, while even in the rural areas it was declining fast, with the introduction after 1878 of primary schools teaching in Spanish (Heiberg 1989: 46). For men in Arana's position none of this represented progress: the challenge was to understand what had been lost, how it had been lost, and how it might be resurrected. He developed a complex vision to understand and respond to these processes, a vision which involved the fusion of Catholicism and the concept of a people, a race.

Catholicism is fundamental to Basque nationalism in all periods. In the 1890s it provided Arana with a powerful interpretation of the evils of contemporary society. The Papal Encyclical '*Rerum Novarum*' of 1891 is the foundation stone of Catholic social doctrine, setting out the theological roots of modern Error and their disastrous social consequences. It denounces the values of liberalism and of the misery and injustices created by capitalism, and denounces those who oppose these injustices through socialism. Both parties to the modern conflict fall into the errors of materialism and Godlessness. Catholic social doctrine advocated a middle path, a more egalitarian society, with widespread property-holding, based in families with their own patriarchal authority, sanctified by God. Small family businesses were the best building blocks of society, to be buttressed by voluntary associations like guilds and co-operatives. Here, and in later elaborations, the autonomous family is realised best amongst the peasantry, whose experience brings them closer to God, since their lives are subject to the laws of nature and not of man.

Basque nationalism included all these elements of Catholic social doctrine: the moral and intellectual leadership of the church; a violent opposition to the liberal state and an even more abiding one to socialism; a denunciation not of industrialism per se but of 'bad' industrialism, with its concentrations of wealth and poverty; the centrality of the peasantry in a moral society. All this is characteristic of emerging forms of European Christian Democracy, and could have been subsumed into it. But there were specific local problems, and a 'loss' which could not be addressed by a pan-Spanish political project. The evils of Bilbao were a direct result of the defeat of the Carlist cause and the *fueros*: a system of local self-government associated with a hard-working, God-fearing peasantry. Regaining these rights from the Spanish state would be a necessary part of the resurrection, while the ills of industrial growth and socialism were linked directly to migration. The concept of the Basques as a distinct people emerged as a second organising concept alongside aspirations for a particular kind of society, and was fused with Catholicism in the developing nationalist ideology.

The notion that the Basques were a separate people was not a given. At the elite level Basques had played important roles in the Spanish church and state for centuries: not just participating as a minority amongst a pre-constituted Spanish majority, but playing a significant part in that process whereby Spain itself had been constituted. There are indications (Conversi 1997: 46 and sources) that Arana and his brother came to this nationalist conception not as something obvious, but initially as a troublesome deduction from the fact that they wanted for the Basque provinces something (the foral regime) which the rest of Spain had rejected. At first Arana's energies were concentrated on the problems of Vizcaya province and the case for its independence; then they broadened to include all the Basque provinces. Even when the concept of a Basque nation had been consolidated ideologically, there remained the constitutional question of whether the rights of the Basque people could be realised within a Spanish federation or required independence. Amongst Arana's early supporters, and throughout the next 100 years, regional and separatist aspirations co-existed within the nationalist movement.

The Basque people constituted a race: a group defined in the first instance by biological descent. Race as a concept is diachronic, requiring an historical time-frame: the identification of ancestors who had the same characteristics as those found today, who had reproduced themselves by marriage within the group, and not introduced 'impurities' by marrying out. Once the premises are accepted, the arguments linking identity, purity and continuity become self-confirming, and tautological. Scientific conceptions of race and culture were very widespread in nineteenth-century European intellectual circles, and were picked up in nationalist movements, but Arana could also draw on sources embedded in local thought and practice. The historic status of 'collective nobility' granted to the Basque people had at its heart all the essential connections: an unconquered homeland, pure blood lines and hence uncontaminated Catholic faith. Under the foral regime, noble status had been used in certain periods to prevent the settlement of migrants. Local naming and inheritance practices continued to stress continuity and unbroken bloodlines. Biological research on blood groups and skull shapes, and philological research on the antiquity of Euskera, provided a discursive framework for local understandings of identity and ancestry.

Of course, positing the existence of a pure Basque race (as opposed to an impure Spanish one) does not tell you immediately who is part of it, and who is not. The movement needed criteria for inclusion or exclusion, and neither residence in the provinces nor language would do this. In the early years membership of the Basque Nationalist Party was made conditional on possessing four Basque surnames, since they demonstrated purity at least to the male great-grandparents, and surnames continued to act as immensely important markers over the following century. However, a political movement needs a collective imagery, a sense of who 'we' are, fleshed out with some social and cultural characteristics. If the only answer is to go down the street asking people their

great-grandparent's surname, we have a rather pedantic and impoverished basis for launching a grand vision. Here again the peasantry came
to the rescue. The urban centres, whether the steel towns or the old *villas*,
could not offer the essential imaginary – they were too heterogeneous,
and full of people who had migrated and intermarried. Only the rural
world of the *basserias* offered an image of continuity and unequivocal
Basqueness: you knew without asking. The peasants themselves
remained largely indifferent to the nationalist movement for another
generation, but they provided the cultural themes which were essential
for its articulation.

I have suggested that the core of Basque nationalism lay in Catholicism
and racism, which fused and subtly altered each other. The main reason
for seeking separate development was to allow Basques to practice their
Catholicism away from the corrupting influence of liberal Spain and
Godless socialism. Nations constituted extended families and had been
created by God; each nation formed its own relationship to God, prayed
to him and communicated with him separately (Heiberg 1989: 53). If
each nation has a unique relationship to God, it is an easy step to seeing
the Basques as having a special relationship to him, being a chosen
people, and elements of such a view can be found.

Arana's religious beliefs were condensed into his notion of *Jaungoikua*, a critical
emblem of the nationalist doctrine. The Basque word for 'God', or, literally, 'Lord
on high', *Jaungoikua*, was the fountain of all sacred and worldly authority inside
the Basque country. But implicitly, *Jaungoikua* was an exclusive God. He was the
God of the Basques and, as such, provided an autonomous framework for
religion. (Heiberg 1989: 54)

There was also theological speculation about the antiquity of Euskera:
was it the language common to mankind before the Tower of Babel,
spoken in the garden of Eden, spoken by the Creator himself? (Conversi
1997: 64). In this environment, even God could be ethnicised. These
ideas represented more extreme tendencies, and should not obscure the
more general point that Catholicism was central to the conception of the
Basque nation, and enriched its cultural definition and celebration, so
that, for example, Easter Sunday, the day of the Resurrection, was chosen
as the Day of the Fatherland, a national holiday.

The relationship between the Basque race and God is a more central
issue even than language and territory, at least in this period. Arana
argued that the patria could not be measured by the territory it occupied,
and would still be Euskeria 'even if it were moved to an island in the
Pacific' (quoted in Heiberg 1989: 52). Similarly, 'if our invaders learnt
Euskera we would have to abandon it ... and dedicate ourselves to

speaking Norwegian or some language unknown to them' (Heiberg 1989: 53). These were rhetorical flourishes, designed to stress the gulf between the Basque people and all others: it did not mean that language and territory were not important – indeed, they would become more so in the twentieth century.

Arana and his colleagues devoted immense energy to documenting and promoting that rural culture which had been lost to the Spanish 'invasion'. The first priority was Euskera, which none of the early nationalists spoke. They learned it, attempted to purify it, wrote grammars and dictionaries, and developed a publishing industry which used it. They founded and co-ordinated circles dedicated to the study of rural culture, and in the early decades of the twentieth century they threw themselves into the organisation of music festivals, dances, oral poetry competitions, games and other folkloristic practices of peasant life. They organised mountaineering clubs, with Sunday excursions into the sacralised mountains, clubs which later became the basis for the armed militias. As Heiberg argues, all this activity did not constitute any simple celebration or resurrection of a threatened rural way of life. It was instead a process, organised largely by and for middle-class urban people, which selected out those aspects of rural life, and those rural attributes, which would act as ethnic markers. That is, they celebrated a set of cultural activities which would provide the lines of inclusion and exclusion in Basque–Spanish relations.

So far we have dealt with the discourse which constructed the Basques as a nation, and the use made of rural culture in the construction of ethnicity by the urban middle class. But delimiting ethnic categories is not the same thing as building a nation, which is achieved by the political work of a movement generated around a discourse. It involves the construction of new lines of social cleavage through exclusionary practice; the interpretation in ethnic terms of selected existing lines of cleavage; and the permeation of the ethnic discourse into the understanding of relations and activities. Of course, much of the work generating ethnic categories – the language revival, the publication industry, the cultural study circles – themselves generated social environments and new systems of communication linking nationalists together. The movement grew in the early twentieth century through the proliferation of new organisations, which took root in a variety of social contexts and linked the Basque provinces.

The Partido Nacionalista Vasco (PNV) was established in 1895 as a federation of community nationalist clubs called *batzokis*, active in both the Basque provinces and amongst the Basque diaspora in the Americas (Heiberg 1989: 70). The PNV established a youth branch in 1901, principally amongst students, and a nationalist labour union in 1911, which was active amongst the white-collar employees and the rural migrants employed in the smaller factories. There was also a very successful Association of Patriotic Women. The movement quickly gained the enduring support of the lower clergy – not surprising given the centrality of Catholicism in the nationalist project, and the attacks on the church from liberalism and socialism. The PNV enabled the clergy to embrace an activist role in relation to Basque rural culture, and begin the work of creating an educational system in Euskera.

These and other organisations formed the 'infrastructure' of the nationalist movement, a sufficiently dense network of organisations 'to create a parallel society in which the *abertzale*, the true Basque, could operate in most spheres of public life closed off from the outside, the anti-Basque world' (Heiberg 1989: 73). Nationalists had to turn race (or ethnic identity) into a politically operative category, implying that descent was a necessary but not sufficient condition for the ascription of 'Basqueness'. True Basques were patriots (*abertzale*) whose political and moral commitment to their race showed through their actions, and led them to live their lives within the 'parallel society' which had been constructed. As we shall see later, this implies that the focus of political activity was not the fight with the Spanish state for independence, so much as a struggle with other political forces for power and hegemony within the Basque provinces. The PNV eventually decided to contest elections, and Arana himself was elected to the Vizcayan provincial government in 1898. There were periods of electoral growth and decline, but by 1930 the PNV had re-emerged as a well-organised third force in the provinces, even if its support was very uneven. It was strong in Bilbao and the rest of Vizcaya, where it came second to the socialist bloc; it was virtually absent in Alava and Navarra outside the provincial capitals.

The early social base of the PNV was created, in Heiberg's terms, out of the fusion of two rather different social groups: an urban manufacturing middle class, pragmatically seeking greater regional autonomy within Spain, and a radicalised petty-bourgeoisie with separatist aspirations. The programmatic differences between these two sectors is not surprising, since manufacturers and shopkeepers tend to have rather different economic horizons. There were feuds and changes in policy, but overall the movement held together, and the existence of two programmes within one party allowed it to widen its base, which included many of the clergy, the intelligentsia, and the professional middle classes. It also gained some ground amongst the urban working class, whose social roots lay in the rural Basque hinterland. Then, from the 1920s, the PNV began to put down roots in the rural hinterland itself – a vital development in the consolidation of the nationalist movement. The next section will analyse the impact of the nationalist movement when it finally penetrated the rural areas, and uses this ethnography to explore the whole question of national consciousness.

Nationalists and the Peasantry

The village of Elgeta studied by Heiberg is on the boundary between Vizcaya and Guizpazcoa, and had about 1,000 inhabitants at the turn of the century, composed almost entirely of farming households. There were just over 100 farms, mostly run by tenants, with rents fixed in kind and paid to landlords, some absentee, some local. It was a largely unmonetised economy, with considerable poverty, but one of substantial cultural homogeneity, since farmers, tenants and their local landlords all shared a rural way of life and spoke Euskera (Heiberg 1989: 72). The key issue in local politics was not 'Basqueness' but the role of landlords: their manipulation of indebtedness to extract surpluses from tenant farmers and to maintain their position as Carlist bosses (*jauntxoch*) at elections to the village councils.

During the First World War a booming armaments industry in a nearby town spread new economic opportunities into the rural hinterland. Wealthier landlords sold off some of their farms to the tenants and invested their capital in industrial enterprises. Employment in factories and commerce increased in the local area, but such jobs required the acquisition of Spanish. The schooling system expanded and there arose a much more widespread pattern of bilingualism, so that different parts of the economy were conducted in different languages. These were also correlated with generation and kinship status, since financing an education (in Spanish) became one of the ways of 'paying off' the non-inheriting heirs of the farmstead. Roads were built and commercial life expanded, creating new occupational categories: shopkeepers, barkeepers, traders, teachers. Throughout the 1920s 'modernity' penetrated the Basque countryside in the form of money, bureaucracy, literacy and Spanish.

Attempts by the PNV to penetrate the rural world prior to 1914 were a failure; in fact, Heiberg's brief comments (1989: 68) suggest a wonderful comedy of manners. Nationalist activists and intellectuals would emerge from the city, dressed in neat imitations of peasant attire, speaking (presumably) a very rudimentary form of Euskera, and totally ignorant of the hard parameters of rural existence. They held meetings to extol the peasants as noble savages, pure custodians of a threatened culture, but

they mostly ended up having lunch with the priest and a few local professionals, who were perhaps the only people they could understand.

A decade later the political climate began to change. The majority of
the rural population wanted their children to have a Spanish education,
jobs in factories, and Spanish-speaking marriage partners. But the
economic changes had created new lines of cleavage within local society.
The eldest son, who inherited the *basseria*, was now locked into a static
and relatively disadvantaged way of life, and often had trouble finding a
wife. The non-inheriting siblings gained an education and employment
in a more prosperous urban environment, and the effect was to reverse
one of the basic tenets of the rural kinship system. The cleavage was
between the rural world of farmsteads and the new urban world of flats,
jobs and education which was emerging in the villages, and which fell
within the orbit of a larger urban and commercial world. Heiberg insists
that the division was one between a Basque rural and a Basque urban
culture, that the population of both worlds had local and Euskera-
speaking origins, and that there was no in-migration at this stage. She
also argues that few people in the local area had the slightest interest in
the relationship between the Basque provinces and the rest of Spain.

The PNV, which continued its proselytising mission during the seven
years of the Primo de Rivera dictatorship, began to make some converts
in the village, and its support became clear when elections were held
again in 1930 – firstly amongst the farmers, the *basseritarak*. The move
from tenancy to ownership had cut many free from patronage ties to the
Carlist landlords, and in addition the PNV had emphasised a rural,
Catholic programme of land-ownership and credit facilities aimed at
improving farmers' welfare. PNV-sponsored cultural activities also
provided new forms of recreation in the village and appealed to local
youth. The nationalist appeal to other sectors was more complex. Many
factory workers and small entrepreneurs had turned to the socialist and
republican parties, which established a nucleus of support in the village.
Other members of Basque urbanised society had their progress blocked
by the ostracism perpetrated against migrants from poor, rural, Euskera-
speaking backgrounds. The PNV found converts in this sector too, and in
the early 1930s constituted the largest political bloc – and the moral
majority – within the village.

The PNV had constructed its nationalist discourse around the figure of
a pure, devout, Euskera-speaking male peasant, inhabiting a
mountainous rural arcadia. When the rural villages were indeed
populated by Catholic, monolingual, Euskera-speaking subsistence
farmers it made few converts. When the penetration of the Spanish state
and capitalism had effected a rupture in that rural continuity and

produced greater economic and cultural heterogeneity; when the pre-eminence of the *basseritarak* had been undermined; when the church for the first time found itself with local enemies in the form of socialism – then the Basque nationalist movement found more fertile ground.

It is not obvious that the PNV would have maintained its hegemony in a local society that was moving so fast along divergent economic and cultural trajectories. The Civil War that erupted in 1936 changed everything. It split the Basque provinces down the middle. Carlist Navarra rose in support of Franco, while the nationalist insurgents and the industrial working class of the coast supported the Republic. The PNV itself was caught in the middle: was it better to support the Catholic but centralising nationalists, or the left, which was Godless but more pragmatic on autonomy? A narrow majority of the PNV sided with the Republic and went down fighting. The reprisals carried out against the losing side in the Basque provinces were brutal, and although perhaps a smaller proportion of the population was killed than in Andalusia, this was combined with very repressive measures against Basque nationalist aspirations. Those threatening the unity of Spain, like the lower Basque clergy, were treated very harshly. The use of Euskera was prohibited in public places, in names and inscriptions. The defacing of gravestones was a disturbing instance of tactics which became more systematic and generalised in Balkan 'ethnic cleansing'. The Franco regime stamped the movement into the ground, but in so doing also provided confirmation that there was a distinctive Basque culture, and that its existence was a threat to Spanish unity. Acts of barbarism, like the bombing of Guernica, confirmed the popular legitimacy of those who upheld the cause.

In the late 1950s and 1960s a renewed and more militant form of Basque nationalism grew up in the seminaries and amongst the student body, and this will be examined in the last part of this chapter. First we need to examine an important and prior theme: the way in which nationalist discourse and practice became embedded in a pattern of social relations, so that everyday life was *self-evidently* a manifestation of a distinctive ethnicity. I shall use Heiberg's ethnography (1989: Chapter 9), looking first at forms of association and community.

Rural social relations featured two forms of voluntary association which created patterns of inclusion and exclusion. The first – *auzoak* – was a loose linkage of neighbouring farmsteads which co-operated in labour exchanges on individual farms, and on collective tasks such as road-building. These linked neighbours also attended each other's weddings, funerals and baptisms, and maintained a private chapel with its patron saint. It was an enduring relationship between households combining principles of autonomy and reciprocity. Hard-working, competent farmers, with a common knowledge of production techniques and landscape, shared labour and resources in a pattern of personalised and balanced reciprocity. The egalitarian ethic co-existed with substantial inequalities, which were transmuted into debt relations and patronage; nevertheless it is apparent that the egalitarian ethic did have some effect on the actions of patrons, and was one of the reasons why they failed to crystallise into a distinct social elite.

The second form of voluntary association was the *caudrillas*, which organised social life outside the sphere of domestic relations and work. They were predominantly male associations, involving 8 to 20 individuals, generally of the same age, marital status and lifestyle, hence sharing a common daily timetable. A *caudrilla* did things together, for example meeting twice each day, for a drink before lunch and supper. It might be formed round a special interest in fishing or hunting; whatever its purpose, its members organised banquets and ate together. It provided a support group for the individual, and organised social and cultural events for the village. The *caudrillas* were structured by inclusion and exclusion, they 'were closed, exclusive social units. Gaining entry into a *caudrilla* once it had solidified was very difficult. The most important element for late membership was gaining the confidence of existing members. And trust was not a commodity bestowed lightly' (Heiberg 1989: 156). Once you were in, there was an egalitarian ethic of immediate and balanced reciprocity, and of trust: information and opinions could be shared and would not flow out of the group.

Having sketched these two important associational forms, it becomes easier to deal with the *pueblo*, a term which, as in other parts of Spain, has two meanings: a place (and all those who live there), and a moral community. It is the gap between these two concepts which leads to the emergence of a particular social and moral map. As a moral entity, the *pueblo* consists of all those who show in their behaviour that they subscribe or adhere to certain values. These may vary over time, but would include those values (equality, honour, prudence, modesty and hard work) which are embedded in the associations. Hence, conceived as a collectivity, the moral *pueblo* consists of all those who demonstrate those values, in the context of the *auzoak* or the *caudrillas*.

The physical map of the *pueblo* is made up of what Heiberg represents as four concentric circles, based on gradations of inclusion and exclusion. At the core is the moral *pueblo*. Outside that, in a second circle, are those who have breached local norms – for example of reciprocity, trust or collective endeavour – and are subject to either temporary isolation or irreversible ostracism. Such people are 'us' – those from inside – but morally lapsed. In the next circle are those from outside, professionals such as teachers and doctors who are resident in the village but come from elsewhere and are not included in circles of labour exchange or long-term sociability. The outer circle consists of those who are not us and are dangerous, a threat to the *pueblo* and its moral life – archetypically the civil guard, and after the Civil War the police informer: *chivato*. Heiberg suggests that *chivatos* share many of the moral attributes of witches (see also a long discussion in Zulaika 1988).

Ethnicity is not a synonym for culture. Ethnicity, in modern usage, is the property of an ethnic group, a people who consider themselves bounded and distinct, on the basis of criteria such as race, religious tradition or language, and who (the circle turns) possess a common culture. Almost any aspect of culture can be incorporated into an ethnic discourse, but the discourse itself is inevitably selective, and does not cover the whole of, or all aspects of, the cultural life of those who form part of the ethnic group. The point becomes clearer if we present the issue the other way round, adapting the perspective of the early nationalists themselves: that there existed a Basque race, and that all those with a certain genealogy (or surname) belonged to it. Basque culture is thus the culture of those who belong to the Basque race. This presents a mind-numbing problem of incoherence. Some members of the Basque race spoke Euskera, some Spanish; some were bankers devoted to high finance, others Jesuits devoted to spiritual exercises; some were farmers with their flutes, some were church-burning leftists – and that is before we consider the American diaspora. So a selection had to be made: some ways of being Basque were purer and more authentic than others, and a narrative was constructed which explains why certain ways of being Basque in the present are truer, because they are continuous with the past. For reasons examined earlier, the choice made by the nationalists was to prioritise the peasantry, and peasant culture and values came to epitomise the Basque race.

Outside the nationalist perspective, there is nothing strikingly distinctive about the social forms of the Basque peasantry described above. The ethnography of rural Europe is full of accounts of labour-exchanging neighbours, drinking circles, hunting parties, distinctions between insiders and outsiders, closed or tacit systems of knowledge and moral communities. In fact, the group of Sussex University graduates organised by Professor Bailey to study European mountain villages, from the Pyrenees to the Austrian Alps, devoted several books to the ethnography of reputation, honour, trust, autonomy and equality (Bailey 1971; 1973). The associational forms may be slightly more 'crystallised' in the Basque villages – and there was the additional issue of diglossia – but in all mountain regions there was a substantial gulf between local patois and the dominant language of the state. But the similarities between Basque and other peasantries were of no interest to urban nationalists – they were not attempting to construct a Spanish peasant movement, let alone a Green International like the politicians of the Balkan agrarian parties between the wars.

When in the 1920s the PNV broke through in the rural districts, they did so by capturing the economic, political and above all moral centre of the *pueblo*. They had developed a programme aimed at the *basseritarak*; they used forms of political organisation and mobilisation which nested into the *caudrillas*, and provided a discourse which interpreted all the key values of the *pueblo* – embedded in everyday social practice – as attributes of the Basque people. In that way everything done by the core 'moral community' of the *pueblo* was an affirmation of Basque ethnicity. Having captured farming households at the core, support for the nationalist movement spread out into the *pueblo* – but not completely. It reinterpreted old and new lines of social cleavage as boundaries of the nation; thus the local landlords with their authoritarian ways were enemies of the Basque nation, as were the emerging factory workers with their Godless–Spanish–socialist beliefs. The internal divisions of local society were essential to the operation of the movement. Finally, we should note that the PNV affirmed the values and importance of this rural way of life precisely when the position of the *basseritarak*, and the central place of practices based on continuity and inheritance, were being subverted by new economic conditions. This is why so many analysts, whether or not they use the conceptual vocabulary of tradition and modernity, link nationalism to the phenomenon of transition.

In the period of reconstruction after the Civil War, and throughout the Franco period, the Basque provinces (except Navarra) suffered cultural repression combined with very rapid economic and demographic growth. A state policy of economic autarky created a second wave of industrialisation which, although fragile in the long term, transformed the previously untouched rural areas, including Alava and Navarra. As a result villages like Elgeta began to experience for the first time the impact of migration from other regions of Spain. Its social infrastructure, built out of long-term patterns of reciprocity, could not absorb migrants, the distinction between 'those from here' and 'those from outside' strengthened, and as migrant numbers grew, so did outright hostility.

In the 1960s and 1970s local society became increasingly segregated. The *caudrillas* tightened and became more exclusive, partly as a response to political oppression. They transformed themselves into *sociedad* (clubs), sought lockable premises where they could talk and eat in private, and turned down all requests by migrants to become members. Some later provided the terrain in which ETA activists and supporters operated. Village bars were largely abandoned by the local population, and were frequented by outsiders. Intermarriage was extremely rare. Migrants tended to find employment in the factories, while local people worked on the farms or industrial co-operatives which were created as a more egalitarian alternative to the factories, on the Basque model of Mondragon. The church organised separate services in Euskera and Spanish. The most significant development was the establishment by a young priest of an *ikastola* – a school funded by private donations – which would teach local children in Euskera. In the 1960s there was considerable scepticism about the educational value of Euskera – by the 1970s it was the self-run school for the *pueblo*, and only migrant children were attending the state school. At that point local society was totally segregated: in work, kinship, socialisation, festivities, worship and education.

The clubs, co-operatives and school had been created and run by the *pueblo* and were seen as 'ours'. At a certain point 'we' became, once more, 'we Basques'. The ground had been laid in an earlier period, and there was an older generation of PNV supporters to keep alive nationalist discourse in the privacy of the clubs. The Falange, and later the church hierarchy, had acted vigorously, and counter-productively, against opposition articulated in terms of the rights of the Basque people. However, Heiberg and her informants in Elgeta are adamant that the polarisation of the village and the reconquering of municipal government were not (at that time) driven by Basque nationalism, and were achieved before it spread into the village. The spread occurred only in the 1970s, when some of the clergy and a group of village youths began to campaign vigorously for the *ikastola* and for folklore festivals. Active support for them became a marker of opposition to the Franco regime and of political commitment to Basque nationalism – both in terms of the social life of the *pueblo*, and the political destiny of the Basque provinces. The moral system of the *pueblo*, and the impermeable barriers it constructed against outsiders, may have pre-existed the spread of a radical nationalist ideology. What the new movement did was confirm these social divisions, express them in purely ethnic terms, and link the politics of the village to the national question of Basque–Spanish relations. It made them homologous: Spaniards out of the Basque country (Euzkardi); Euzkardi out of Spain. In the political environment which emerged after Franco's death in 1975, the radical nationalists triumphed and in villages like Elgeta public life became monolithic. The factories employing migrants shut, partly through lack of credit, while the co-operatives survived; the state school was shut leaving only the *ikastola*. Some local people thought that the result was a happy extended family of moral harmony, others that the social environment was poisonous. As the recession bit in the 1980s, the migrants began to leave.

Conclusion to Part 1

Basque nationalism is articulated around the rights of a people who recognise themselves as having a common culture, one which is distinct from their Spanish or French rulers. A national identity is inevitably an historicised concept; it is a narrative which defines who we are, in part, through reference to ancestry, and an enduring being or essence. Though the narrative can encompass change, there is always a core theme of continuity – of unchanging attributes, qualities or relations. The continuity can be provided by a cultural tradition, by unbroken residence in a territory, or by descent: in narratives of national identity these often overlap, or one element comes to stand for another which is submerged. In the Basque case, for most of the historical period we have been examining, continuity is understood in terms of descent: the nation is constituted by the living representatives of the Basque race.

Anthropological approaches adopt widely different stances to the issues of boundedness and continuity which are found in nationalist movements. For example, Zulaika, an important scholar in this field, can write that 'All prehistoric reconstructions point to the unique occupational centrality of hunting in Basque culture from the Palaeolithic era down to the Christian era' (Zulaika 1988: 187). Here the historical record is confirming continuity, and hence identity. Other approaches are more 'presentist' in suggesting that a nationalist narrative constructs an identity by selecting out cultural themes, or by the invention of tradition. These approaches have their own conceptual difficulties, in that they tend to accept 'ethnicity' as a useful term for the analysis of cultural difference, but may dissociate themselves from the premises on which local constructions of ethnicity (or nationhood) are based: the existence of a people who share a common, bounded and enduring culture. It is the connection which the term ethnicity makes between 'a people' and 'a culture' which creates difficulties within anthropology.

This chapter has suggested that in the Basque provinces there is evidence of substantial continuities in the way in which some people gained their livelihoods and in certain cultural forms, including language; though these economic and cultural practices became salient when they were no longer taken for granted, but simply one way amongst others. However there is a difference between evidence for cultural continuities and arguments that 'Basque culture' (something which is the property of a Basque ethnic group) is a distinctive and historically stable entity. Nor obviously do these cultural continuities imply that relations between the Basque and the Spanish people are invariant over the last 500 years, let alone since the Palaeolithic. The interesting issue is how cultural differences become ethnic ones.

Basque nationalism emerged in a particular social milieu – amongst the intellectuals and sections of the middle class in the rapidly industrialising towns of Vizcaya at the end of the nineteenth century. I have tried to show how key aspects of this political project were shaped by the experience of these social actors in a specific environment: the defeat of Carlism and the attacks on the church; the rise to power of a liberal oligarchy; the dislocation of industrial growth; mass immigration and the very rapid consolidation of a socialist movement. Squeezed between the political forces organising labour and those of the major industrial interests, the middle strata developed a movement which would make their own position, cultural values and resources once again central within this transforming social world. That meant the position of hardworking, industrious, autonomous, property-owning, devoutly Catholic, *local* people. It also meant that they created a narrative of loss, and a movement which would attempt to reverse the process of destruction.

The resulting movement successfully mobilised the political energies of a growing proportion of the population, though not without tensions and paradoxes. Its discourse was a fusion of Catholicism and 'commonsense' views of race and descent, which constructed the Basque nation as an historic people. The nation was exemplified in the present by the peasantry, but others belonged by birth and could identify with it through their actions – by participation in the cultural and political activity of the movement itself, with its festivities, clubs and sections. I have suggested that if we treat the culture of the Basque peasantry as forms of knowledge and practice embedded in economic and social relations, we find that it shared many features with other European peasantries. While the nationalist movement initially celebrated those linguistic and 'folkloric' features of peasant culture which would demarcate a nation, it also seized on those cultural items and performances which would resonate outside as well as inside a peasant milieu. The activists were not generating class or peasant politics, but stressing vertical lines of inclusion and exclusion – the shared culture of 'those from here'.

We can say that the movement worked to create a cross-class alliance, but this is a rather thin description. More than that, it was the creation of a network of communication and activities which brought together people of different social backgrounds in an environment where they shared a common language, understandings, and a commitment to a political cause. The movement consolidated when it became embedded in social relations, and the nationalist discourse provided a frame of reference for the interpretation of everyday social relations. At this point, when a group of farmers in Elgeta meet for a drink and to swap stories, they are manifesting their ethnicity: a minor example, perhaps, but a revealing one when compared with rural scenes elsewhere. In effect the discourse had fused with the existing moral map of a *pueblo*, and provided a language for social relations and experiences. As it did so, the sense of home and homeland was particularly valorised (see Zulaika on the *basseria* and 'The House of my Father', 1988: 131). This is the emotional charge of a nationalist movement: that it articulates those personal and intimate relations and experiences (what some have called the primordial loyalties) often strongly linked to a sense of threatened continuity. The articulation is re-focused around an ethnicised sense of self.

The movement operates through inclusion and exclusion, selecting out cultural differences and social divides, and re-inscribes them as the qualities of a people. This is the ethnicisation of social relations and, operating simultaneously with the opposition of other political forces and the repressive policies of the Spanish state, it builds the Basque nation. At the end of it the nationalists have, for better or worse, hegemony over what constitutes Basque culture. In the rural district studied by Heiberg the social divides articulated by nationalism varied over time. In the 1920s the divide was between 'true Basques' and those local people who were either Carlist patrons or socialists; in the 1960s it was between local people and migrants. The fact that the nationalist movement operated through the articulation of oppositions meant that it was not even necessary for it to create a standardised Basque culture (though that was one of its aspirations); it was only necessary for it to articulate, from a morally central space, an opposition between 'us' and 'them' in cultural terms which people could recognise in their various local realities. It worked through homology, not homogeneity.

This brings us to the last point, about the homology between Spaniards in Euzkardi and Euzkardi in Spain. Basque nationalism always had a dual agenda. Its dominant ideology was about the rights of the Basque people in relation to the Spanish state, either for greater autonomy or for independence. Its second objective was political control within the Basque provinces, in competition with a strong, immigrant-backed socialist movement. Heiberg argues that amongst local supporters the issue of independence was always less important than control within the Basque provinces, and this makes Basque nationalism rather unusual. The point may be valid if comparison is made with the early twentieth-century nationalist movements in the multi-ethnic empires of eastern Europe, but it seems to me that most west European regional and nationalist movements have had this dual character. They have opposed the power and influence of those local people who support, or are agents of, economic and political forces which are represented as alien, and gained strength by assimilating the question of foreign influence to the claim for national autonomy.

PART TWO: ETA, THE STATE AND VIOLENCE

Industrial growth in the Basque provinces – especially Vizcaya – at the end of the nineteenth century generated two political movements: Spanish socialism and Basque nationalism. One hundred years later a more complex political situation existed. A second, more diffuse wave of industrialisation began in the late 1950s, bringing factories and migrants even to the more rural districts, where two very different kinds of society were uneasily juxtaposed. It did raise living standards, so that by the 1970s Alava province had the highest per capita income in the country (Ben-Ami 1992: 148); but it also brought dislocation and social problems: pollution, poor housing and inadequate social services (Clark 1984: 200ff.). Moreover, the industrial boom barely lasted a generation, since the manufacturing base was unable to compete successfully in an international market when the protectionist policies of the Franco era gave way to European Union accession. By the 1980s there was recession, a rise in unemployment to 20 per cent (far higher amongst young men) and the reversal of previous migratory trends, with a net outflow of population. Farming may have continued to play an important role in nationalist imagery, but it now only employed 10 per cent of the population (Zulaika 1988: 102).

Cultural and political oppression co-existed with these economic difficulties. Children from Euskera-speaking households were pitched into monolingual Spanish schools; the display of symbols and cultural forms associated with Basque nationalism was suppressed by Franco's police in ways which were both petty and brutal. When ETA emerged in this environment in 1959, and later developed a military strategy, the repressive policies of the state intensified, and the Basque provinces became the most-policed region of the most-policed country in western Europe (Clark 1984: 263). The police used mass arrests and the torture of detainees to crush ETA. An international outcry over the trial of one group at Burgos in 1970 saved their lives, but others were executed right up until Franco's death in 1975. During the democratic transition, and later under a socialist government, the violence escalated, with ETA killing a wide range of targets and the government employing death squads to eliminate activists.

The combination of economic, ethnic and democratic struggles created a very complex political landscape after 1960. Forms of class action re-emerged in the last decade of the Franco regime – there was even a wave of strike action in 1962 – under the hegemony of various pan-Spanish left-wing parties. The PNV continued to articulate a 'conservative' version of Basque nationalism, first from exile in France, then with increasing success within the Basque provinces after the return of democracy. The third element was the radical nationalism of ETA, a clandestine organisation dedicated to armed insurgency and the con-struction of an independent Marxist state of the seven Basque provinces. Another way of representing this shifting political landscape would be to say that it was a nationalist project, split between a conservative and a radical wing, and a socialist project split between a Spanish and a Basque component. Either way ETA (and the political parties which were spun off from it, or represented it in democratic arenas), became the radical centre, reshaping the discourse of Basque identity and condi-tioning the action of all other political actors in the region.

ETA (Euzkardi and Freedom) emerged in 1959 out of a variety of youth groups circulating in the ambit of the PNV. Information on the social background of its members is incomplete and contested, but Clark (1984: 143–52) suggests that in the late 1970s ETA had over 1,000 active members, over 90 per cent of whom were men in their middle- to late 20s, recruited predominantly from Euskera-speaking areas and from families with a history of nationalist sympathies. They came from a variety of backgrounds: manual and white-collar workers in the smaller factories, the self-employed, students, priests. However, this clandestine movement relied on the support of a variety of people, and operated in an environment with a pre-existing structure of closed associations – the *caudrillas* described above.

These young people were convinced of the need to mobilise in the face of the political stasis of the Spanish regime and the perceived inertia and compromises of the PNV. Conversi (1997: 254) notes that since 1973 ETA has operated under the slogan 'Actions Unite, Words Divide' and taken an anti-intellectual turn, so that debates become almost a sign of weakness, producing so many reasons for not acting. Commitment to the cause was evaluated in terms of action, with death in violent con-frontations with the Spanish state representing the ultimate sacrifice. If action took precedence over theory and programmes as a mode of political mobilisation, there was nevertheless a theory of action, made explicit in the writings of Krutwig (a key ETA leader). Armed attacks on the state would provoke brutal repressive measures against civilians, unleashing a spiral of violence which would render state power naked, and generate increased support for the armed struggle. A similar strategy was found in other armed insurgency movements of the 1970s, including the Italian Red Brigades. Wieviorka (1993) has argued that the violence of the ETA movement is a way of binding together the otherwise incompatible political objectives which run through it. In analysing what would constitute the liberation of the Basque people, most commentators have identified a plurality of positions, and acknowledged that there are continuities as well as ruptures between ETA and the PNV.

The first strand in the ETA movement is the struggle for the cultural rights of the Basque people and for national self-determination. The dominant definition of who constitutes the people is 'those who speak Euskera'; indeed a Basque who fails to use the language is a traitor to the cause (Zulaika and Douglass 1996: 166). In important ways this represents a break with the racist doctrine of the Arana period, opening up the possibility – denied by racism – that a person can become Basque and assimilate through acquiring the language. Urla (1993) has argued that the battle over the standardisation of the language represented a modernising project, and an acknowledgment that, especially since the 1980s, Basque culture was dynamic and open to new forms. If in one sense the definition of the nation was more open, it was still incorporated in an exclusionary view of local society: the preservation of Euskera required an independent state, and the rejection of a policy of bilingualism. The switch from a racial to a cultural definition of the nation does not represent a sudden or complete break with earlier political discourses. Most early ETA activists were formed in a milieu of passionate Catholicism, of the village school (*ikastola*) and an attachment to rural lifestyles and imagery: although some currents of the movement denounced traditionalist versions of Basque culture, they remained potent. Racist formulations and the denigration of immigrants resurfaced here and there in the nationalist movement, but more important than the formulations was the environment in which it operated. After all, ETA did not construct its category of the Basque nation in a vacuum, but in a society where an earlier nationalist movement and state repression had generated deep social divides and entrenched patterns of exclusion and inclusion. Social practice is more important than shifts in discourse.

The second strand in ETA viewed the primary struggle of the Basque people to be against capitalism, reflecting the more general renewal of Marxist political culture in western Europe in the 1960s. The primary locus of the struggle was the factories, and the enemies included those

'Basque' oligarchs whom the PNV had denounced at the end of the nineteenth century. At various times ETA intervened in labour disputes by kidnapping or killing industrialists whose workers were on strike. But there were tensions within this 'red separatism'. ETA activists were not well represented in the major factory complexes, where the working class was primarily made up of 'immigrants' and organised – first covertly and then overtly – by Spanish left-wing parties. The actions of this faction of ETA, and their declarations of sympathy with other oppressed groups in Spain, 'seemed calculated to alienate the Catholic small-town milieu which had been the traditional base of Basque nationalism' (Sullivan 1988: 96).

The third strand within ETA were known as the 'Third-Worldists' (Clark 1984: 34). The relationship between the Basque provinces and the rest of Spain was conceived as a colonial relationship, and the struggle against it to be comparable to those taking place in 1960s Algeria, Cuba or Vietnam. The difficulty with this analogy arose from the fact that the Basque provinces were the richest region of Spain, and that their representatives had played a leading role in the formation of Spanish economic policy for a century. So the analogy required some flexibility in the use of categories (the Spanish, the Basques), but it had the virtue of moral simplicity, and provided a framework for combining arguments about economic and cultural forms of oppression. It was a convincing argument for as long as the Spanish government maintained a strong military presence in the region, and ETA's own strategy assured this.

These three strands coexisted uneasily within ETA, with the balance shifting over time. An assembly held in the late 1960s adopted the conception of the *Pueblo Trabajador Vasco* (the Basque working people), defined as those who earned their living in the Basque country and who supported Basque aspirations. Sullivan (1988: 55; 62) suggests that the conception allowed the different factions to interpret the protagonists of the movement in different ways: the '*Pueblo*' could be essentially Basque or essentially working class. However, the history of ETA is one of repeated fission and a shifting kaleidoscope of breakaway groupings, between ETA 5 and ETA 6, ETA military and ETA political. Rather than summarising the chronology of the factions (there are excellent accounts in Clark 1984 and Sullivan 1988) we can indicate a synthesising principle: each time a group of activists begins to privilege the class struggle over the nationalist one, they tend to form alliances with other extra-parliamentary or par-liamentary left-wing forces. They are then ejected or marginalised by ETA as '*españolistas*' – pro-Spanish. A whole series of groupings, including the moderate nationalist party Euskadito Ezkerra (EE), have followed that trajectory. Over the long run the dominant discourse within ETA became that of the 'Third-Worldists', which combined class and nationalism within an anti-colonial framework.

The use of violence was itself one of the factors which precipitated the different factions within ETA. Targeting the police and the military may be a coherent strategy for a national liberation movement operating within a highly oppressive regime. But for those who believed that they were engaged essentially in a class struggle, requiring mass mobilisation, forged in factories and city neighbourhoods, this kind of violence built no solidarities, and could indeed be alienating and counter-productive. The shift towards class therefore tended to mark not just 'pro-Spanish' sympathies but a gradual renunciation of armed insurgency. Once again the 'Third-Worldists' held the ring, embracing a Che Guevara-style strategy of freedom fighters taking to the mountains to liberate the people from freedom and colonialism. It was a strategy built on a very slippery concept of 'the people', on the prospect of escalating guerrilla warfare (unlikely given the dense militarisation of the state), and on a 'pressure cooker' model of revolutionary violence: screw them down and they will blow. For a sustained critique of explicit and implicit uses of pressure cookers and volcanoes as models for political violence, see Aya (1990).

The second important factor in understanding the strategy of violence and the tensions within ETA is the wider political context. Dictatorial regimes do indeed fuse political struggles, at least to the extent that they are responsible for both economic and cultural forms of oppression, and no progress is possible on any front until the regime is ended. The assassination of Admiral Blanco, Franco's chosen successor, in a bomb attack in Madrid in 1973 cemented ETA's reputation as the force that was doing the most to end the dictatorship, and suffering the greatest losses in the process. Spain's transition to democracy after 1975 is usually seen as one of the quickest and most successful in European history, culminating in the 1978 Constitution which gave greater autonomy to the Spanish Basque provinces than the statute of 1936. But the late 1970s – which might have witnessed a reduction in tension, a separation of political projects and a return to democratic strategies – were marked by an escalation of violence, and the consolidation of ETA and its strategy of armed insurgency.

In practice, no transition is that quick or complete. Even democratic parties take time to coalesce and establish their social bases, and you cannot put an insurgency movement on hold for five years while the picture clarifies. As importantly, there were clearly continuities during the change of regime. There were no significant purges of state personnel, and the frequently brutal institutional culture of the police and the military in the provinces remained. There were strong forces within the state opposed to any weakening of Spanish unity, and delays in the process of devolution even after the passing of the new constitution. The PNV emerged as the largest party in the region and, after a period of tactical manœuvres, developed an accommodation with the government, accepted the overall framework of autonomy within Spain, and settled into regional power. A minority continued to demand independence and pursued it through violence. Activists raised finance through revolutionary taxation (or extortion) from business people, raided deposits for explosives and reactivated the killing campaign: 15 died in 1975, 64 in 1978 (Ben-Ami 1992: 159). They killed members of the civil guard, high-ranking military officers, police informers and businessmen, extending the campaign to Madrid and the tourist resorts. By June 1993 they had killed 749 people, 453 from the security forces and 296 civilians (Zulaika and Douglass 1996: 194).

The political developments in the Basque country have stimulated a substantial volume of research and analysis on the tragic levels of violence produced by the conflict between an armed insurgency movement and the state in a western European democracy. They have also traced the evolution of a nationalist movement which, though originally built around a conservative and racist discourse, developed a radical class-based project. Wieviorka (1993) has argued that there is in fact a connection between the levels of violence and the attempt to hold together disparate political projects in one movement. We can look at each of the issues in turn.

A democratic government has a more difficult task in dealing with armed insurgency than a dictatorship. It normally employs two strategies – one 'political', one repressive – with a complex interaction between the two. The first strand is negotiation, attempting to convert demands into formulae which can be dealt with democratically, and in the process splitting away those in the movement whose demands can be accommodated within an existing or acceptable constitutional arrangement. This is assisted by raising the stakes for those who remain outside the negotiating process. The strategy aims to reduce the size and popular support for the movement down to a small core, whose demands are not negotiable and whose actions are categorised as criminal – a problem of law and order. But in the face of entrenched clandestine activists, normal policing – that is, the identification and conviction of criminals – is comparatively ineffective. Intelligence-gathering is hard, informants (*chivatos*) extremely vulnerable, and a process of law with evidence and witnesses may become virtually inoperable. If armed insurgents can act with impunity there is a crisis for the state, and its agents may adopt other methods: arbitrary raids and mass arrests (14,000 since 1978), torture, confessions extracted under duress, the suspension of civil liberties, military tribunals, imprisonment without trial, forced exile on suspicion of subversive activities. State violence can take many forms, and most of them have been employed in the Basque provinces. The most ruthless involved death squads, a dirty war of covert operations by militarily trained agents from within Spain, and from the underworld of western European right-wing extremism. These anti-terrorist forces (GAL) operated with the knowledge and connivance of the socialist government in the 1980s; 67 are thought to have been killed, in addition to those who died in overt police operations or while held in custody (Zulaika and Douglass 1996: 205).

Ruthless and illegal forms of state violence themselves represent a realisation of ETA's own strategy and, far from isolating intransigent activists, may increase their support. If the state uses illegal and violent strategies to maintain its power within the territory, it undermines its own authority and the distinction between legal and criminal activity. The two contending forces become comparable, lending legitimacy to the political claims and methods of radical nationalists.

The spiral of violence is generated both by state practice and factors internal to radical nationalism. Clark (1984: 278) has suggested that 'ETA is caught in the grip of forces that are created by insurgent violence itself, and that tend to make the violence self-sustaining.' He suggests that exiting from the organisation can be difficult, partly because it is seen as a betrayal, with a real possibility of feuds and revenge killings. Heiberg (1989) and Zulaika (1988), in very different ways, reveal the dynamics of life within ETA – of small groups of young people, who have normally shared large parts of their adolescent and young adult life in the same *caudrilla*, and whose active political life has only been possible through the maintenance of very strong relationships of trust and loyalty. The actions of the movement have generated a cumulative narrative of heroism and sacrifice which gives sense and value to those of the next generation. Armed insurgency movements are totalising, and the price of exit is very high in terms of a break with the bounded social world within which their lives are lived, and the rejection of the discourse within which they have meaning. Indirect evidence for this comes from those who have successfully left. Wieviorka (1993: Chapter 13) conducted interviews with a group of veterans of the armed struggle, whose comments on those who stayed inside demonstrate a total rejection: 'they are crazy, paranoid' (183); 'mafiosi who live on extorted money', 'nazis' (192); 'with a taste for necrophilia and a cult of the dead' (185).

Wieviorka advances an interpretation of the continued violence which is more structural and political. ETA and the umbrella organisation of the MLNV (Basque Liberation) pursue disparate political objectives (speaking for the suppressed nation, social movements and the revolution) and attempt to forge them into the 'myth of an all-Basque movement', with the ETA activists as a vanguard elite. In argument and confrontation with representatives of any single struggle, ETA claims both that these lesser objectives are part of a higher good which can only be realised through separatism, and that its activists are the only ones seriously committed to its realisation. In a series of dialogues (Wieviorka 1993: Chapter 14) the separatists position themselves very clearly within local political culture, castigating representatives of the PNV as conservative and indifferent to working-class issues, and castigating the Socialist Party as indifferent to the needs of the Basque people. This positioning would of course be destroyed if independence were ever achieved. Violence – as well as being a response to state violence, and a strategy chosen to achieve separatism – is a way of creating solidarity within the movement and fusing together its objectives. A marked ambivalence emerges in Wieviorka's dialogues with the veterans, who feel that in renouncing violence they have also renounced any hope for a significant transformation of society. The interviews evoke a loss of certainty and a nostalgia for a time when a radically different future was an orienting narrative, similar to those of the class movements dealt with in earlier chapters.

The ethnographic accounts make clear that, for the activists of the separatist movement, there is indeed only one issue; that at the level of experience the movement and its strategy of armed insurgency have fused what, elsewhere and for others, were distinct political struggles. I shall conclude this chapter with some brief comments on the more general implications of this situation.

The Basque separatist movement is not the only one in contemporary Europe which is attempting to articulate both a nationalist and a class-based programme – there are many others, including the republican movement in Northern Ireland and the Occitan movement in southern France. In each case there is recourse to a generalised notion of 'the people' which is capable of interpretation in both a class and an ethnic sense. We shall be returning to the dilemmas of 'populism' in the Conclusion. In the Basque case there is a specific problem. In those periods during the last 100 years when the working class of the Basque provinces had been politically constructed, and the temporary absence of dictatorship had allowed them to mobilise, this had been achieved by Spanish left-wing parties, primarily amongst an 'immigrant' population.

The Basque nationalist movement had been, in part, a reaction to this development, and elements of exclusionary practice still cast a long shadow over it. Whatever the general problems of combining class and ethnicity, it is certainly hard to construct a persuasive case for an independent socialist state if the movement conceptualises the majority of the working class as belonging to another nation, and cannot make their experience and aspirations central to a future society. In this context it is worth noting that, although we know a great deal about rural loss and the rich symbolism which frames the life and death of radical nation- alists, we know much less (from English-language sources) about the double oppression experienced by a migrant worker in Bilbao in 1890 or in Alava in 1970.

But these exclusionary practices are not the whole story, and within the movement some factions have encouraged changes in both discourse and practice. Race and fixity give way to process; as many authors have noted, you are not born a patriot but become one through action, and that road is increasingly open to all. There are verbs for becoming a patriot (*abertzaletu*), and a word for the adopted homeland (Conversi 1997: 252). The identity narrative is increasingly constructed around the nationalist movement itself, with its own continuities and forking paths, while the original cultural content becomes considerably diluted and attenuated. Increasing numbers of young people from non-Euskera- speaking backgrounds participate in the movement, though because of regional devolution and educational reforms they are also more likely to acquire some knowledge of Euskera and to use it in a variety of contem- porary media and events.

The depth of this support is shown by the vote for Herri Batasuna (Popular Unity). This party and its successor (which rarely takes its seats in the regional assembly) is part of the liberation movement, often speaks for ETA in public, and has been described by political scientists as a catch- all anti-system party. In the last decade it has attracted up to 20 per cent of the electorate in the provinces, and provided a home for those engaged in a variety of struggles – against the nuclear industry, and in favour of environmental issues or gay rights (though less amongst feminists). The focus for direct action has been the Spanish state and its local agents; its cutting edge – implicitly and explicitly – has been those who are prepared to use violence to achieve a new nation-state within a European federation. If this is a nationalist movement, it is not now one which is generated principally out of a struggle for cultural rights (around language, for example), but one which feeds off a variety of acute social problems, not least those derived from crippling unemployment and recession. It has to be said that adopting a more open set of criteria by which to judge somebody a patriot does not necessarily lead to a changed view of patriotism, or represent tolerance for cultural diversity.

SOURCES

The literature in English on Basque nationalism is very substantial, with major contributions from anthropologists, historians and political scientists. The best and most recent work also draws extensively on Spanish scholarship. Although the material is generally of very high quality, it is rather compartmentalised, with little attempt to synthesise the different lines of analysis or interpretation which have emerged.

This chapter draws most heavily on Marianne Heiberg's *The Making of the Basque Nation* (1989), which covers the economic and political transformations of the Basque provinces, but crucially also contains the kind of ethnographic detail which is essential for thinking through the ethnicisation of social relations, and the way this political movement came to articulate core areas of experience. In addition to the monograph, in the first part of the chapter I have also drawn on Heiberg (1975 and 1980), and a number of other scholars for ideas and information: Ben-Ami (1992), which is an excellent short overview; Conversi (1997), who provides a wealth of political history in a comparative perspective; Clark (1979); Greenwood (1976 and 1977); MacClancey (1993); Payne (1975); Urla (1993) and Zulaika (1988). After outlining my ideas on a narrative of loss to Diego Muro-Ruiz, he supplemented my limited knowledge of Spanish by recommending Juaristi (1998), where similar themes are explored and the parallels with Ireland developed.

In the second part of the chapter, concerned with ETA and violence, the key source is Wieviorka (1993). The author collaborated with Tourrain in the development of action sociology; whatever reservations one may have about research method, it has generated very stimulating data and insights into political process. I have also gleaned much information on political developments from Clark (1984), Sullivan (1988), Laitin (1995), and, on local constructions, from Zulaika and Douglass (1996).

7 YUGOSLAVIA: MAKING WAR

A study of nationalism in eastern Europe inevitably has a different agenda from one of a society such as Spain. There are important and profound differences in the timing of industrialisation, the way empires broke up, the character of the ruling groups in the new states, the organisation of cultural diversity, the experience of communism. All these will emerge in the discussion of Yugoslavia, and give breadth to the overall treatment of nationalist movements. At the same time we shall return to themes which have already been explored, including the reaction of rural populations to the repeated disruptions and dislocations of the twentieth century, and the role of violence in political action. Above all we shall return to the forging of national identity in the interaction between local sets of social and cultural relations and state-level politics.

Yugoslavia has long been a theatre of war: a territory which was periodically contested by two great empires, then devastated in two world wars, and again in the 1990s. Nobody born in this region in the last two centuries could have reached the age of 45 without being caught up in war and most would have been involved much younger, and more than once in their lives. They were caught up not as civilians in a society conducting a campaign against some remote enemy, but in a massively destructive land war which passed through their homes. Between 1912 and 1918 one-quarter of the population of Serbia died, including two-thirds of all men between 15 and 55 (Judah 1997: 101). Yugoslavia suffered as much death and destruction as any society in the Second World War (Simic 1973: 59). In these traumatic periods, as in the uprisings and liberation struggles of the nineteenth century, war was followed by massive internal migrations.

A second important feature of Yugoslav society is that, like the rest of the Balkans, it was until recently overwhelmingly rural. Nearly 80 per cent of the population were employed in agriculture in 1920; and even in 1961, after a decade of very rapid industrial growth, agriculture still employed half the workforce (see Ramet 1996: 75; Simic 1973: 30). The proportion was always higher in Serbia. It was an economy built on

131

intensive subsistence agriculture, but if peasants are to achieve a livelihood for themselves and their children they need residential and social stability.

The co-existence of these two dimensions suggests important tensions within Yugoslav historical experience. It had a time-scale of a rural society, privileging continuity and stability in the relationship between kin groups and land; but there was another of destruction, migration and wars, which can be retrospectively collapsed into each other as episodes in one epic struggle. One reason for introducing these dimensions at the beginning of the chapter is that they represent the two axes of an identity narrative (as suggested in the Introduction) – one establishing roots and continuity, the other an enduring opposition. They are also both important factors in understanding the power of nationalist movements in the region, since national identities draw heavily on peasant culture, while war is one of the ways in which nationalism is consolidated. This chapter will analyse the dominance of nationalist politics in post-communist Yugoslavia, not as the return of ancient conflicts between peoples, but as specific political movements operating on complex lines of economic and cultural differentiation.

Many accounts of the Yugoslav tragedies of the 1990s begin with a history of various named peoples – Serbs, Croats – who were first conquered and then struggled to achieve their independence from the Ottoman and Austro-Hungarian empires. There is much to be learnt from accounts which chart a millennium of ethnic history, but in the absence of other social and political themes, of the kind which had been important in earlier scholarship, they lead to only one conclusion: we are witnessing the latest stage in an ancient Balkan tribal conflict. It is helpful to read not just recent studies, in which every historical conflict 'prefigures' the current violence, but also those which bring out the complexity of economic and cultural differentiation. There are certainly continuities within Yugoslavia, in terms of social practice and of the historical narratives which interpret the conflicts. But we should also see the territory of Yugoslavia as an area of constant flux, of evolving cultures and shifting social boundaries, interspersed with periods when those boundaries harden into total exclusion. In the twentieth century the most important cause (not consequence) of that hardening has been war.

The second reason for introducing these other dimensions is to emphasise the variety of political movements in the Balkans, and that they have economic as well as cultural roots. The peasantry was at the centre of many political struggles. In the nineteenth century this region had become a granary for Europe, exporting cereals, but this was only achieved through very high levels of extraction and falling living standards for the rural population. The Russian Revolution in 1917 sent shock waves through the Balkans. It was a major factor stimulating a massive land-reform programme, a measure designed to stave off more violent upheavals. The reforms broke up the large estates, distributed land to the peasantry and strengthened an intensive farming system. Not all the problems of rural overpopulation and malnutrition were solved, but the 'Green Rising' and the land reform was acclaimed by one observer as 'a vast victory for the peasants, and therefore a vast defeat for the Communists and the capitalists' (Chesterton, 1922, in Mitrany 1951: 118).

Politically, the main beneficiary of the new measures that enfranchised the peasants of eastern Europe was not the Communist Party (with its Red Peasant International founded in Moscow in 1923), since it had little support until the Second World War. Instead it was the Agrarian or Populist Parties, which themselves organised a co-ordinating Green International in Prague throughout the inter-war years, attempting to develop common programmes across national divides, both within existing states and across international borders. Co-operation was possible, and began to take practical forms: such practice was considered treasonable by those in power (Mitrany 1951: 137), and eventually contributed to the downfall of the Agrarian Parties. Their programmes had been built around democracy, property rights, the establishment of co-operatives, small-scale rural-based industrialisation, direct proportional taxation, and increased public expenditure in rural areas.

These radical programmes brought peasant organisations into conflict with urban mercantilism (Mitrany 1951: 121). If the vast majority of the population was rural, the towns were inhabited predominantly by traders, bureaucrats and soldiers. Belgrade, the capital of the new Kingdom of the Serbs, Croats and Slovenes, created in 1918, was subsequently a city 'top-heavy with petty bureaucrats and corrupt officialdom' (Simic 1973: 59). The city was populated by those who benefited most from the new state machinery, financed from revenues extracted overwhelmingly from the rural population. The clash between rural and urban interests took a brutal turn in many parts of the Balkans. In Bulgaria the Agrarian government was overthrown, and its leader Stamboliski murdered in 1923 (Bell 1977). Radic, the leader of the Croatian peasant party, was assassinated in 1928, and this was followed in 1929 by the establishment of a dictatorship in Yugoslavia under the king. Throughout eastern Europe (with the exception of Czechoslovakia) the peasant parties lost power to the officer class.

Accounts of war in Yugoslavia in the 1990s stress, quite rightly, the legacy of the vicious conflicts between 1941 and 1945, but by taking a longer time-span we get a rather more complex picture. We see, for example, the enduring importance of the social division between town and country, and the way it frequently cut across emerging national identities (see Halpern and Halpern 1972: 12). Peasant culture has been the bedrock on which Serbian and Croatian nationalism were built, but this does not necessarily mean that the peasants themselves were always nationalistic. When in 1868 Serbian landlords began replacing the Ottomans they were considered just as rapacious as their predecessors, and faced continuous uprisings from their serfs (Judah 1997: 54). In the land-reform period after 1918, there was a general 'hardening of the nationalist temper' where the social division between landlords and peasants was overlaid by linguistic and religious differences (as in Croatia and the Vojvodina), but the same political agenda was articulated by the peasants in other areas where this was not the case (Mitrany 1951: 88). The dissatisfactions articulated by nationalist movements need contextualising in relation to both peasant–urban relations and central state policy. Put another way, if we insist on reading the country's political history only in terms of nationalism, we also have to admit that there is more to national conflicts than the conflict between nations.

This chapter will concentrate on the period since the Second World War, with only selective forays into earlier periods. It starts with a brief summary of communist economic development, followed by a section on the organisation of the Yugoslav federation, which recognised both territorial divisions (republics) and peoples (*narod*). The central section is an account of local-level diversity, drawing out the connections between social groups and cultural differences, and the way these were transformed by communist practice. Ethnographic analysis is vital to our understanding of nationalism, but has to work with details which are not generalisable to the whole of Yugoslavia. I have chosen Bosnia (Serbia would have been simpler, Macedonia still more complex) partly because of the quality of the ethnography, partly because so much was at stake there in terms of cultural pluralism and international intervention. The last part of the chapter deals with the destruction of Yugoslavia, covering the strategies of the dominant groups under late communism, the principles which informed international intervention, and the connections between ethnicity and violence.

YUGOSLAVIA UNDER COMMUNISM

In 1941 Yugoslavia was overrun by Italian and German troops. Fighting and reprisals continued right up to and beyond the liberation of Belgrade in 1945. During the war a pro-Nazi government – the Ustashe – was established in Greater Croatia, while further south, Serb nationalist forces – the Chetniks – led by officers of the Yugoslav army, continued to fight for the construction of a greater Serbia. In between were the partisans, led by Tito and built around a nucleus of a few thousand communist activists who had survived the first onslaught in 1941. They dug in deep in the mountains of Bosnia, which was never totally controlled by the German army, and expanded into a large and successful liberation army – a process aided by the switch of British support from the Chetniks to the partisans. It was a long war, with brutality on all sides. Over a million people died, including tens of thousands of civilians who were rounded up in death camps, the most notorious of which was run by the Croat Ustashe at Jasenovac.

Yugoslavia emerged after 1945 as a Stalinist state, centralised around its Communist Party and the army, dedicated to 'Brotherhood and Unity', the construction of Yugoslav citizens and comrades. There were further massive internal migrations – for example from the poorer mountain regions to repopulate the Vojvodina, from which German populations had been evicted. There was ruthless suppression of internal dissent even after the break with Stalin in 1948, and the later emergence of Yugoslavia as a champion of the Nonaligned Movement. Communist economic policy was directed towards rapid industrialisation, achieved as in the Soviet Union through the control of labour, the extraction of raw materials, and by holding down agricultural prices. In the early post-war period there had been a land-collectivisation programme, quickly reversed, and agriculture remained dominated by small property-holders and co-operatives. Very high levels of investment in heavy industry produced some of the highest growth rates in Europe up until the mid-1960s, and the cities boomed. This was not a standard 'command economy' since market principles were introduced into the supply of consumer goods, and there were measures of 'workers' self-management' in the factories, but the all-important decisions on capital investment were made on a country-wide basis and remained under central state control.

From the 1960s various social and economic problems with the chosen development path began to emerge. There were growing debt problems, and rising unemployment, to which partial solutions were found by opening up the country to mass tourism, and allowing Yugoslav citizens to travel abroad for work (creating a million Yugoslav *gastarbeiter* by 1980). Central planning of capital investment could deliver a period of rapid growth in terms of infrastructure and basic industrial goods, but it suffered from rigidities which became more serious in the 1970s and 1980s, with changing consumption patterns and an evolving global division of labour. It was easy to go on building large steel plants and refineries; harder to adopt new technologies and flexible production patterns. The communist government had taken over a country with very marked regional disparities, but their own policies exacerbated the situation, so that by 1970 the per capita GDP of Slovenia was six times that of Kosovo. A struggle emerged between the richer regions (Slovenia, Croatia) which wanted more decentralisation and market-based decision-making, and the poorer regions (Serbia and the south) which wanted centralisation and redistribution (Bojicic 1996).

Regional disparities – and nationalist stirrings in Croatia which had been suppressed in 1971 – fed into a substantial constitutional reform enacted under Tito in 1974. In a complex series of moves, very substantial powers were devolved to the regions, while the federal government retained control over foreign policy and the military. At the same time the Communist Party (now organised as a Yugoslav League) strengthened its powers within the new structures, vetting and co-opting officials in all key positions (Dyker 1996: 55). The 1974 constitutional reform is referred to by all analysts of the break-up of Yugoslavia, partly because it is the moment when the current political units, which had had a shadowy existence since 1918, emerged with substantial autonomy; and partly because the federal constitution itself was both weak, in that the power given to the republics impeded federal decision-making, and rigid, in that it made further reform – including the introduction of democracy – virtually impossible.

The republics gained presidents, parliaments and ministers, ran education policies, decided how many Slovene-speaking doctors were needed, ran their own media. Not least, the republics had control over their economic policy, and there was a shift from nationwide planning to regional autarky. In other words the bosses of each republic had as their main concern

to obtain sufficient resources to finance their own favourite development projects, aimed at as high a degree of self-sufficiency as possible ... in the case of the less-developed regions this frequently meant going for prestige projects, offering very little productive return for their regional economies. (Bojicic 1996: 43)

Policies directed at the efficiency of the Yugoslav economy as a whole were marginalised, and the more industrially developed republics traded increasingly with other countries rather than with other republics, while being constrained to transfer funds to the poorer regions. The statistics documenting this process are complex and contested, because they are an important part of the argument about the inevitability of the break-up of Yugoslavia and the future of its successor states. What we can note is that the 1974 Constitution generated very strong regionally focused power bases within each republic, and that opposition to communism within Yugoslavia took a distinctive form. Whereas elsewhere in eastern Europe dissident movements had grown up opposing totalitarian state power, in Yugoslavia the enemy was the federal authority which was limiting republican autonomy, while organising opposition across republics was extremely difficult (Vejvoda 1996a).

The reforms did not halt the decline of the Yugoslav economy. In fact, in conjunction with the worsening international climate, they accelerated the centrifugal processes and the country's economic collapse. In earlier periods the Yugoslav federal government had been extremely skilled in manipulating its international position in the highly polarised world of the Cold War:

> Its independence was in fact a strategic resource that depended on the conviction of Western powers that national communism in the Balkans was a propaganda asset, and that Yugoslav neutrality could be a vital element of Nato's strategy of containment in the east. (Woodward 1996: 157)

A kind of soft deal was done. Yugoslavia acted as a buffer zone against Soviet threats in south-east Europe, and denied the USSR access to the Adriatic. In return Yugoslavia obtained western loans, and access to western markets and technology, not least to build up a very strong military capability. By 1980, with European recession hitting both exports and the demand for migrant workers, Yugoslavia had accumulated $20 billion of foreign debt and was struggling to deal with the financial crisis.

In 1982 Yugoslavia, like other debtor countries, was the subject of an IMF austerity and stabilisation programme. It involved restrictions on imports, a freeze on wages and salaries (while inflation continued), cuts in public expenditure and an inevitable sharp rise in unemployment. The result was high levels of hardship and economic insecurity, and it is worth dwelling briefly on the impact of this for rural–urban relations. Accounts of urbanisation throw up one interesting statistic: even in the expansionary period of the 1970s, urban families spent 70 per cent of their income on food (Simic 1973: 115). Ethnography reveals that an important strategy for piecing together a livelihood in town was to maintain direct access to rural production, either through retaining property rights or through extended kin links (Despalatovic 1993). In the 1980s the government programme for absorbing surplus labour included sending 'the actual and potential unemployed back to families, villages and private-sector agriculture and trades' (Woodward 1996: 159) – a difficult policy given the generally high educational qualifications of such people. Radosevic (1996: 72) notes that by 1989 nearly a quarter of the population were living in poverty, but that half of these lived in the countryside: 'this was a prime reason for the absence of social unrest, as ownership of land permitted the rural population to be self-sufficient in food'. Olsen (1993) also notes increased dependence on personal networks in order to obtain a livelihood. We do not have many ethnographic accounts of the poverty and frustration in the 1980s. However, there is evidence that many people who had embarked on urban-based livelihoods in industry and the professions found themselves forced back into total or partial dependence on peasant agriculture and extended rural-based kin.

Towards the end of the 1980s there was one last attempt to reassert some central control over decision-making and deal with the financial problems. In 1990 the last federal prime minister, Markovic, tried with some success to enact a stabilisation package, and regain control over the money supply. But the balance of power had shifted decisively to the republics, and Milosevic effectively subverted these controls and asserted the right of bosses in Serbia and elsewhere to print money in order to deal with regional deficits (for the Serbian Great Bank Robbery see Dyker 1996: 59). In addition the international context had been utterly transformed. The Cold War was over, and Yugoslavia no longer had the same strategic importance as a buffer zone. Eastern Europe was filling up with countries anxious for European Union accession and western credits: Yugoslavia went to the back of the queue. Some international players were blind or indifferent to the fate of Yugoslavia, while others had already begun unilateral negotiations with its constituent republics, on the premiss that they were dealing with nascent nation-states. We can now turn to the issue of nationalism within Yugoslavia.

REPUBLICS AND PEOPLES

The social organisation of cultural diversity is crucial in understanding the political and military struggles of the 1990s. Analysis involves unpacking categories like Serb, Croat and Muslim, and looking at the ways in which these identities emerge over time, and at different levels of social organisation. I shall start with the categories used by the state, which are relatively simple, and then move to an ethnographically-based discussion of identities in everyday social life, with their more complex and fluid practices of inclusion and exclusion.

The 1974 Constitution had created the republics as 'sovereign states' within the federation: bounded and increasingly differentiated territorial units with their own economic and social policies. Citizenship of Serbia thus denoted a civic status. However, the Constitution also reaffirmed that the Yugoslav federation was made up of peoples or nationalities, and the right to secede from the federation was granted to peoples. There were six such peoples (or *narod*): Slovenes, Croats, Serbs, Montenegrins, Macedonians and the 'ethnic Muslims of Bosnia' – although the latter were only granted this status in 1971, previously having been a religious category. These were the 'constituent' peoples of Yugoslavia, in the sense that the majority of such people lived there. The state classification of cultural diversity, however, was more complex than this, in that there were other peoples (termed *narodnosti*), such as Albanians or Hungarians, who were generally distinguished from *narod* in that the majority of the population lived outside Yugoslavia (Sorabji 1989: 13). There was a further category of minorities such as Gypsies. These are census categories, and there was one other: that of Yugoslav, used as a default for those who declared national categories not recognised as one of the six *narod*, and more positively by those who identified with the federation.

What happened after 1974 was that the interests of the republics as 'sovereign states' were increasingly merged with those of the *narod*, who represented the majority, so regional elites articulated their interests in terms of nationalism. Instead of consolidating civic conceptions of citizenship and developing policies of cultural pluralism at a society-wide level, the issue of cultural difference was cast in territorial terms, and considered solved by the creation of homelands. Yugoslavia became a patchwork of majorities and minorities – but particularly minorities: those who did not live in the republic of their *narod*, or who had the misfortune to live in a republic which had no overall majority – the ill-fated Bosnia.

Yugoslavia illustrates many of the general problems of establishing democratic political life in post-communist states, and the rise of nationalism as the principle legitimating ideology. The problems became exacerbated because the regime had not only suppressed democracy until the day of its collapse, it had created a political structure which channelled all dissent into arguments about the rights of national majorities and minorities. The political logic which follows is nicely summed up in the question, 'Why should we be a minority in your state when you can be a minority in ours?' (quoted in Vejvoda 1996b: 260). Nationality became central in the political life of the republics, and their fortunes incorporated into historical narratives of their constituent nations. This nationalist reading of Yugoslav history becomes so pervasive that it inhibits thinking in any other way about the kinds of economic and cultural differences which predated the 1990s descent into war. Accounts of those wars (and of Yugoslav society) are predicated on the assumption that there exists a finite number of stable ethnic groups

(6, 18 or 43) which, as it were, 'pre-exist' any political process, and themselves create a political problem, since their co-existence within one state is increasingly seen as problematic. The solution – the key to creating stability – is conceived in territorial form, with the creation of new nation-states – or ethnically homogeneous cantons, as in Bosnia. The consensus around this solution emerges from a macabre dance between the 'international community' and a particularly ruthless post-communist generation of leaders within Yugoslavia. There *is* now separation into 'mono-ethnic' blocs, and deep social divides do indeed exist, but this situation has been the product of political processes – some old, some very recent – which have reworked and simplified complex and fluid patterns of cultural differentiation, and overridden those social and political divides which did not involve ethnicity. We will get a better grasp of that process by looking at some ethnographic accounts of social life from periods when war did not appear inevitable.

CULTURAL DIVERSITY IN BOSNIA

In the early 1970s Lockwood studied a district of small towns and villages in Bosnia, Skoplje Polje, and one element in his account of social organ-isation is the *nacija*. He says that the term corresponds to 'nation' and refers to this aspect of social organisation as ethnicity, a usage discussed at the end. He says that each *nacija* is largely endogamous, that

members of each consider themselves, and are considered by others, as a distinct and unique variety of mankind, are perceived as possessing certain God-given qualities regarded (albeit wrongly) as unchanging and unchangeable. A local folktale illustrates this well. Each of the *nacije* went to God and one by one were given their special attributes and way of life by Him; last of all came the Gypsies, and therefore they must be satisfied with the leavings of others. (Lockwood 1975: 22–3)

Given the nature of the categories, it ought to be a very straightforward matter establishing how many *nacije* existed locally. The first answer is that six were traditionally established in the district: Serbs, Croats, Muslims, Gypsies, Cincars and Sephardic Jews, until these latter were killed in the war. However, a series of post-war migrations have brought in a further 9 categories of people, making about 14. The first observation is that this small area shows considerable heterogeneity, and that through massacre and migration the composition changes significantly with every generation.

A second point to note, following Lockwood, is that though local people consider these categories to be stable and God-given, this is unlikely to be the case. The Cincars, for example, are said to be 'serbianised Vlachs'.

Once herders, then settled coppersmiths, they have switched their language (from one that was Latin-based to Serbo-Croat) and are now assimilated into the Serbian population whose religion they share. If this was an 'ethnic' category, it was defined as much around a now-defunct occupational niche as by cultural distinctiveness, and no longer has any salience. It is refreshing to see some verbs ('serbianisation', 'magyarisation') breaking through the language of timeless categories.

Thirdly, how clear-cut are these categories in local usage? What in fact makes a social grouping a *nacija?* For example, the Gypsy population, who declared themselves Muslims in the census, are internally divided on almost every count. One group – the whites – are permanently settled, are almost all metalworkers, and speak only Serbo-Croat. The others are nomadic, occupy a different economic niche, and speak Romany amongst themselves. There is no intermarriage between the groups (Malcolm 1996: 116). How many ethnic categories is that, and for whom? A more important example is the Muslims. The designation covers categories of people who, both historically and in contemporary society, are sharply distinguished. Some are descendants of a Muslim aristocracy (*begovi*), some are peasants; they each practice different forms of Islam, maintain different lifestyles, and do not intermarry. According to Lockwood (1975: 29), in the local folktale, 'They go separately to receive their God-given attributes. And *begovi* actually did share certain characteristics with *nacije* as determined by ethnicity. Social boundaries of the group were as sharp as with any ethnic group.' Are these two Muslim ethnic groups, or one ethnic group divided on class lines? What has happened to the simple, unambiguous, ethnic categories?

In order to understand phenomena such as *nacije,* and avoid naturalising them, we need to open up the whole field of cultural differences and how these relate to social divisions. Lockwood's ethnography provides the basis for a fuller picture and shows that in this district cultural variation was present in many domains, from religion and language to cuisine and vernacular architecture. Each of these was seen locally as a constituent part of *nacija* identity; however none of them were sufficient on their own, as cultural items, 'to make the difference', and none of them operated exclusively in the field of ethnicity. Religion was very important, but there was no direct correspondence between religion and *nacija* in the local system, since more than one *nacija* practised Orthodox Christianity and more than one was Muslim. Even in the state system of *narod* classification, there were not always simple correspondences, since there were people who were Muslim by nationality, and Jehovah's Witness by religion (Banac 1996: 146). As for language, 'there is no association of the major dialectical divisions (of Serbo-Croat) with ethnicity in Skoplje

Polje'(Lockwood 1975: 53). Variations in costume, perhaps the most conspicuous everyday markers of difference, in fact operate along several axes. For example, 'The traditional men's costume in Skoplje Polje is nearly identical for all three groups. It differs only in the colour of the sash ... the trim ... and the headgear' (Lockwood 1975: 49). In the next valley everybody wore a different costume, with minor variations in trim for each *nacija*. Dress codes simultaneously carry regional, gender and '*nacija*' markers – only the context determines which will be seen. Early in the twentieth century visitors to Sarajevo found great difficulty in distinguishing between Christians and Muslims because they both wore 'oriental' dress (Malcolm 1996: 167). In diet, there are certain food items which distinguish Muslims from Christians, but overall, 'Like many aspects of culture, Turkish influence on eating habits follows a rural–urban dichotomy rather than Moslem–Christian' (Lockwood 1975: 52). To which we can add, there are many other fields of culture, in the wider sense, where the divide between peasants and townspeople would be much more marked than the distinctions between peasants.

Nacija as a concept in Bosnia, like 'ethnic group' in anthropology, links together a people and a culture. I have drawn attention to the many axes of cultural difference in rural Bosnia in order to make two points. Firstly, we need an approach which keeps in sight those social and cultural processes which will go unrecognised within the dominant discourse of ethnicity. Secondly, in the analysis of ethnicity and nationalism it is a mistake to start with cultural difference, abstracted bits and pieces of folklore, linguistic or religious practice, and attempt to map them onto social groups. Instead we need a different strategy, especially for a society such as Yugoslavia, which has developed more than one way of categorising peoples and cultures. The complexity of this situation will remain a source of confusion if we fail to distinguish between these systems of categorisation, and then smooth the social science terminology of ethnicity over the top. The premature use of the term ethnicity simultaneously adds another layer of labelling, and obscures crucial processes and distinctions in the society that we are trying to understand. For example, the question, 'How many ethnic groups were there in Yugoslavia?' (like the question, 'How many castes are there in India?') is unanswerable unless transmuted into an inquiry about a specific level of differentiation – and even then the answer will depend on who is asked.

I have stressed the existence of at least two levels or systems of differentiation. The first is the pan-Yugoslav categories regulating relations involving state administration and practice: these are few in number and fixed throughout the territory. The second is those embedded in everyday local life – *nacija* in Lockwood's ethnography. The term may not be used throughout Bosnia, let alone the whole of Yugoslavia, but the social patterns it refers to are common: ramifying networks of people who intermarry and consider they share some essential cultural practices, especially involving religion. The number of such groups will vary between localities and over time – they are relatively fluid. Analysing local systems involves contextualising them in the full range of cultural differentiation and in relation to social practice: marriage patterns, economic relations, domestic life, ritual celebrations. It is only through such an analysis that we can gain an understanding of the reasons why certain cultural differences become important, how social divisions are reproduced or disappear over time, and for that matter why certain commonalities (these people are all farmers and speak the same language) may be unmarked or invisible.

The last and most difficult step is to analyse the interaction between these local-level systems and those employed by the state, such as the six *narod* of Tito's Yugoslavia. This interaction takes many forms. Under communism some policies sharpened the social divide between country and city, attenuating other cultural differences; other policies regenerated nationalism. For much of the time local systems may be largely self-referential – 'We have always done things this way around here' – a set of cultural traditions, some derived from religious teaching, some of entirely local significance, 'the origins of which are frequently unknown to those who practise them' (Sorbaji, quoted in Malcolm 1996: 222). But in times of nationalist polarisation, local practice draws on wider referents and, as we shall see, political movements draw heavily on local and domestic imagery. What emerges are the kinds of identity narrative referred to in earlier chapters: the positioning of collectivities within society and within history. For a Bosnian village this involves Catholicism, Orthodoxy, Islam, and the changing position of their adherents within Yugoslavia and globally. The wider reference can subtly transform the local perspective and practice: long-term neighbours, with their overlapping cultural repertoires and village loyalties, become accidentally co-resident representatives of distinct global religious communities.

The reader may object that, having insisted on the importance of distinguishing between levels, I now want to treat them as inseparable. It is worth clarifying what is at issue. The mistake lies in assuming that there is one system of cultural differentiation – which we call ethnicity – manifested both at the local level in a given pattern of social interaction and exclusion, and at the state level in the categories used in the census, the federal constitution or in administrative practice. If we assume that these two levels simply mirror each other, we have written politics out of the picture.

A valuable illustration of what is at stake comes from Sorabji (1989: 38), who, in a critique of Lockwood's use of the term 'ethnicity', was one of the first to stress political processes – the way 'national belonging' provides a new language of claims, rights and interests. There has always been great cultural diversity in Bosnia – what was new from the nineteenth century onwards was the way these cultural groups formed the basis for nationalist politics. In 1878 Bosnia was acquired by the Austro-Hungarian Empire during a period of renewed contestation with the Ottoman Empire. The Bosnian population at that time had three main components: a large number of Muslim, Serbo-Croat-speaking peasants; an even larger number of Christian Serbo-Croat-speaking serfs; and a small elite of Muslim landlords for whom the latter worked, and who used Ottoman Turkish in official communications and in the *sharia* courts (Pinson 1996a: 104–5). In the years following Austrian annexation the strategy of the Muslim elite was to promote organisations which would forge unity across the class divides amongst Muslims, to petition the Austrian authorities over conscription, to champion the legal and religious rights of Muslims and, by the early twentieth century, to form a Muslim political party. The result was a hardening of the boundaries between Christians and Muslims, and the political fusion of the Muslim population, who for the first time found themselves subjects of a Christian government. The Bosnian parliament became divided on 'national' lines, despite the Austrian authorities' earlier hopes for the emergence of a Bosnian national identity which would encompass religious differences.

The crucial aspect of this 'politicisation of difference' is the way in which local-level social relations and 'systems of difference' are transformed and simplified by their incorporation into a state-level political movement. The discrete sets of relationships within villages and neighbourhoods were aggregated together, 'homologised' in a discourse which constructed society as made up of separate (and sometimes antagonistic) nations. Ethnogenesis, and ethnicity more generally, are best reserved as terms to describe this politicisation of difference, and the interaction between local and state systems. This kind of transformation tends to be cumulative, and the differences harden; but this is not inevitable, and in the twentieth century there are processes leading to the softening of boundaries, as well as periods of cataclysmic polarisation. The alternative viewpoint sees ethnic groups as given, and then seeking political expression; it shades into the extreme view that Balkan history is a saga of tribes fighting over territory. Since one of the objectives of recent wars has been to make local and state systems congruent through 'ethnic cleansing', the use of terms which collapse these levels at the beginning of an analysis is a poor basis for understanding the politics of this period.

Analysis of these political processes has to include the history of categories such as Serb, Croat and Muslim; they vary considerably over time. For example, the Austrian policy of promoting a Bosnian national identity was in part directed at countering the claims that the Muslims were really Serbs or Croats – claims which could form the basis for territorial annexation. Even in the early years of communist rule, the Muslims were still expected to declare themselves as Croat or Serb by nationality, since Muslim was not yet a *narod* category in the Yugoslav state. The historical paradox of this is that the Bosnian Muslims could not have been Croat or Serb before they converted because 'no such distinct categories existed in Bosnia in the period before Islamicisation' (Malcolm 1996: 199–200). The use of the category 'Croat' is itself complex. Bringa (1993) has suggested that within Bosnia the term was not used for local Catholics, but referred to inhabitants of Croatia. In Croatia itself, the term Croat includes the descendants of Hungarian, German and many other predominantly Catholic settlers (Despalatovic 1993; Malcolm 1996). The overall history of state policy shows the consolidation of religious categorisation in legal and economic practice from the Ottoman millet system onwards, followed by a nineteenth-century shift such that religious categories became national categories. But it was a shift which left ambiguity, as we see both in the case of Catholics and in the 1971 definition of a new nationality: the 'ethnic Muslims of Bosnia'.

What I have referred to as a local-level system had religious affiliation at its core, though not everybody in Bosnia who shared the same religion was considered to be part of the same *nacija*. A *nacija* was a group of people who were largely endogamous, gave their children a distinctive set of names, celebrated the same religious festivities and *rites de passage*, shared some distinctive features of diet and dress, domestic organisation and popular culture, who resided together and who organised labour exchanges. Most of a rural person's life was conducted with people of the same *nacija*, and when they moved outside this world, to the market for example, they expected their affiliation to be clear to strangers (Lockwood 1975: 49). They engaged with people from different *nacija* when they went to market, worked in the lumber industry, the local factory or the farmers' co-operative, enrolled in higher education, joined the Communist Party, or became labour migrants in Germany. A dry list, but a revealing contrast, and we shall come back to it.

There were two overlapping sets of social relations within which this identity was generated. First the domestic sphere, where certain cultural forms (names, rites of passage, diet) were substantially permeated by religious practice and could not be combined in the event of a 'mixed' marriage. The expectation that the domestic world is uniform, and that 'you marry your own', reproduced over time, meant that kin groups and descent lines belonged to the same *nacija*. The second context was that of neighbourhoods – the relatively bounded world of reciprocity, labour exchanges, borrowing; the world of the informal or the non-monetised economy; and also that of reputations, of trust (and betrayal). In the latter context it was very similar to forms of village patriotism documented from other parts of Europe: Tuscan patriarchs used to say 'take women and [buy] oxen from your own place'. It is a world where people prefer 'to deal with their own', or with 'our people'. There is a mutually reinforcing process operating: your people are those you trust, you trust only your people.

Examining the social context of ethnic identity clarifies the ways in which it is shaped by specific local conditions and by general social transformations. As economic differentiation and social mobility increase, the importance of these divisions – indeed their continued existence – is open to question. They are not incompatible with city life, as becomes clear from Sorabji's (1989: 67) description of a Muslim neighbourhood in Sarajevo. Nevertheless, in the 1980s more than three million of Yugoslavia's 22 million inhabitants lived in 'mixed' households (Woodward 1995: 36), which means that, given the generation lag and the strength of the countryside, there must have been a tendency for them to break down in an urban environment. The industrialisation programme initiated by the communist regime was like a great motor generating mobility and creating urban environments where the question of baptismal names and when to celebrate Easter were becoming less important than other kinds of identity and lifestyle. But this process was checked and reversed. Already, by the late 1960s, migratory flows were becoming more internal to each republic, and in some cases this would lead to greater urban homogeneity. Denich (1993; 2000: 44) has some interesting comments on how Belgrade seemed to evolve from a cosmopolitan federal capital to a Serb capital between her visits in 1972 and 1987. The more dramatic slide in living standards as the economy collapsed in the 1980s created widespread economic insecurity and blocked aspirations. Rising unemployment, the erosion of welfare benefits, the pressure on young people to return to the village and the increased dependence on subsistence agriculture took many people back to a rural world, and to a generalised dependence on kin, patronage links and an informal economy which had been structured around *nacija* identities and divisions.

Yugoslavia was a highly complex society – in the ways in which cultural diversity was manifested, in the kinds of antagonisms that emerged, and in the ways in which identity was constructed (for example in the relative importance of descent). I have suggested that in all cases we need to distinguish between state and local levels of organisation, and examine their interaction (see also Bougarel 1996: 98, 104). The local level, in Bosnia, involves *nacija*: elsewhere in Yugoslavia the term itself may not be used or may have different connotations from those reported by Lockwood, Bringa and Sorabji (Mursic 2000: 68), but we very commonly find named kin and neighbourhood groups which have similar sociological and cultural characteristics. In Bosnia, *nacija* are those who do things together, in a context where there are many ways of doing things. For many the tolerance of plurality in Bosnia made it the place where the Yugoslav ideals were most fully realised (Weine 2000: 404). Some intellectuals and politicians thought that such heightened and reflexive tension – one that had not sought resolution in either segregation or assimilation – was also a model of multiculturalism for the rest of Europe.

Prior to the wars the degree of local mixing varied greatly, as did patterns of localised hostilities. Yugoslav towns in general were hetero-geneous; by contrast in much of rural Yugoslavia villages tended to have only one *nacija* (to be 'monoethnic'). Even in Bosnia, where many villages had two *nacije*, in the 1970s feuds and hostility normally occurred between those of the same *nacija*, or in the rivalry between villages of the same *nacija*. Bax's description of a lawless part of Herzegovina reveals every kind of feud, not least between co-religionists (Bax 1995; 2000). For the national picture, one well-informed observer reckoned that 'Interethnic hostility is no more highly developed in Yugoslavia than in most parts of the world and a good deal less than in some' (Lockwood 1975: 28). We should not portray rural Bosnia as permanently harmonious, since conflicts did occur, and were remembered. It is enough to insist that antagonism was not natural or inevitable, since if cultural differences automatically generated hostility and violence, every metropolis in the world would be a smoking ruin.

The wars of the 1990s were not produced by spontaneous outbreaks of hatred, but by political strategies. The Granada television film *We are all neighbours* is a disturbing documentary on the effect of that war on a Bosnian village where there had been harmonious social relations between Catholics and Muslims. A young woman who worked in a factory which recruited from many villages remarked in an interview that she hated those who spoke of 'our people' meaning 'our nationality'. A few weeks later everything had been torn apart.

DISINTEGRATION

In 1990 the Yugoslav state collapsed. The federal government ceased to control the army (the JNA) so that the various armed forces, the territorial defence units and the police were operating independently. The Central Bank ceased to have any control over financial policy, debt or the currency. Yugoslavia's internal borders became contested and militarised zones, while many of its international borders became porous to illegal trade. The rules of law and citizenship which had regulated the society for 45 years evaporated.

In their place came republican governments, led by people who had made careers in the Communist Party and the state bureaucracies, the *apparatchiki*, who had either broken with the party to construct their own nationalist movements, like Tudjman in Croatia, or had stayed within it, like Milosevic in Serbia. They were linked through patronage ties to managers in the media, industry and finance. Milosevic, who always faced substantial internal opposition within Serbia, denounced the 'unacceptable face of communism', its bureaucratic apparatus, and then at the end of the Cold War moved all the regional assets of the Communist Party into his newly-created Serbian Socialist Party. His most destabilising move was the claim that Kosovo, where 90 per cent of the population were ethnic Albanians, was the heartland of the Serb people. Support was mobilised through a series of rallies at the Field of Blackbirds, the site of a battle between Serbian and Turkish armies in 1389. In 1990 the Serbian government unilaterally abrogated the constitutional rights of the autonomous provinces of Kosovo and Vojvodina and incorporated them into the Serbian Republic. The other main power base in communist Yugoslavia had been the army; proud and well resourced, it was the fourth largest army in Europe, but lost coherence in the transition. The majority of its officers were Serbs and the subject of manipulation by the government in Serbia before 1991. In the wars it shed equipment to Slovenia and Croatia, was purged four times (Vasic 1996) and, as a much reduced force, came under the control of the Serbian government.

In 1990 elections were held in the republics, and these confirmed the strength of nationalist politics – though scarcely unanimously, since Tudjman only got 42 per cent of the vote. It is one thing to ask people who they want to run a republican government, but another to ask them whether they want to abolish Yugoslavia. Most observers agree that in

1990 there was an overwhelming majority for the continued existence of Yugoslavia in a federal or confederal form, but no pan-Yugoslav elections were ever held. In the summer of 1991 Slovenia announced its secession from Yugoslavia: it was the smallest and wealthiest republic, with a strong, exporting industrial sector and international borders with Italy and Austria. It was Catholic and claimed 'ethnic' homogeneity (though 25 per cent of its workforce came from Bosnia and Kosovo). What could be more reasonable than that it extricate itself from a collapsing communist regime, and from the backward southern regions, and realise its European destiny as a nation-state? A brief military engagement with the Yugoslav National Army left 50 dead and Slovenia free, and able to use its own currency, which had been secretly printed in Austria.

Croatia announced its independence at the same time, and both countries were granted recognition by various European powers. However, the military consequences in Croatia were far more serious. Croatia had a substantial minority of Serbs, and the Belgrade government had announced that if Croatia wished to leave the federation, its Serbian population had the right to remain. Some of these people were fully integrated into professional life in the cities, though they soon found themselves disenfranchised. Others were rural people who had settled centuries earlier on the 1,000-mile militarised frontier between the Habsburg and Ottoman Empires, and maintained a strong military tradition. Fighting in these frontier areas involved the army and irregular militias; it included the destruction of Vukovar on the eastern border, the systematic killing and expulsion of civilian populations from strategic areas; rape, torture, and detention camps, which became death camps. United Nations troops were sent in to maintain ceasefires in 1992.

Following a US initiative, Bosnia was recognised as an independent state in April 1992, but its admission to the UN more or less coincided with its destruction. There was never much chance that it would escape the catastrophe. It was highly integrated into the Yugoslav economy and, since 1938, had been chosen as the strategic site for military industry – the 'unconquerable' heartland of Yugoslavia, containing between 60 and 80 per cent of the army's physical assets (Woodward 1995: 259). Croatia and Serbia had decided on its dismemberment amongst themselves, and the war was well prepared. Sarajevo was besieged; Mostar, that other symbol of a now unacceptable 'multi-ethnic' society, suffered constant artillery bombardment. Despite the presence of UN troops and endless ceasefires, war and terror broke up the 'leopard skin' of ethnic co-residence in Bosnia-Herzegovina. The fighting was a three-sided contest (Muslim, Croat and Serb) and involved regular armies, militias and mercenaries. In 1995 the Muslim forces, in provisional alliance with Croatian forces, pushed back Serb forces from much of the north of Bosnia, creating a new wave of (Serb) refugees. The Croats were receiving funding for equipment from the diaspora community, and spent $5.5 billion on arms and on assistance from retired members of the US military (Glenny 1996: 283; Kaldor 1999: 46). It was at this point that the US exerted its greatest pressure for a negotiated settlement, the latest of a long line of maps partitioning Bosnia into separate territorial entities was drawn up, and the Dayton Agreement was signed in October 1995. A UN presence remained in the area and a central government was created on paper, but its presence was highly limited.

Between 1992 and 1995 around 260,000 people were killed in Bosnia-Herzegovina (Kaldor 1999: 31), and 3.5 million Yugoslavs became refugees within Yugoslavia itself, or scattered across western Europe. At the end of the fighting, much of the economy was in ruins, and living standards had plummeted, except for the military. Wars – fought to break up a federal state which was considered 'artificial' and usher in a bright new world of clean-cut nation-states – left behind an even more artificial patchwork of crippled states, para-states, protectorates and lawless enclaves; states which rely on their neighbours, or international agencies, or both, to provide rudimentary policing and much of their defence force; states which have virtually no system of taxation in operation, whose currencies do not circulate; lands where the Deutschmark is the only money that counts. States with a public sector, a rule of law and normal patterns of accumulation can be rebuilt, and in places this is happening; but they are expensive, and since there are those – both inside and outside Yugoslavia – who prosper in the absence of a state, there is nothing inevitable about the rebuilding process. The war in Kosovo and fighting in Macedonia in 1999 were just the more visible parts of continuing instability.

In order to make some sense of this devastation I shall look at two issues in the concluding sections of this chapter. The first is the interaction between the government forces within Yugoslavia and what is called, for shorthand, the international community. The second is the strategy of these same political forces within the republics: how they consolidated their own positions and redistributed wealth within society by the calculated use of violence to polarise social relations on ethnic lines.

INTERNATIONAL ACTIONS

In the speeches of western leaders and in the media, the dominant representation of events in Yugoslavia has been that violence was a direct product of ethnic hatred, which had deep historical roots and exploded after the collapse of communism. Western initiatives were directed towards containing the fighting and providing a framework to help these quarrelsome people to negotiate a settlement. A variant on this representation was that the violence was caused by the aggression of the Serbs, led by Slobodan Milosevic, one of history's psychopaths. These accounts rest on cod history ('some things never change') and cod psychology ('war is produced by aggression'). They tell us something about how western leaders set about obtaining domestic consensus for their policies; they tell us little about the social and political processes in Yugoslavia, or about the way in which the United States and the European Union were instrumental in shaping the political forces which emerged with the breakdown of communism. The increasingly nationalistic governments of the republics developed their political and their military strategies in relation to direct western intervention and interpretation of the west's intentions.

Various scholars have contested the dominant representations of these processes and drawn attention to western complicity in the disaster: the most detailed and merciless is Susan Woodward's *Balkan Tragedy* (1995). At the end of the Cold War, with the west supporting any kind of anti-communism and the European Union expanding, some of the ruling forces emerging in a decentralised and crisis-ridden Yugoslavia wanted independence for their republics and access to western credits. The strategy to achieve this was to argue that the republics represented nations which had the right to self-determination. The Yugoslav constitution, though highly flawed, addressed the plurality of the existing society and created many layers of rights: those of Yugoslav citizens, those of republics, and those of nations and minorities. International organisations which encouraged the break-up of Yugoslavia on the grounds of national self-determination were opening up two immensely dangerous and intractable problems. The first was that of how many nations there were in Yugoslavia. In legal terms the six *narod* were the constitutive nations of a state which it was proposed should be abolished. The status of one of them (the ethnic Muslims of Bosnia) was extremely complex, while there were many other non-constitutive peoples – Hungarians, Albanians, Gypsies – whose rights in the new order would need to be defined. The second explosive issue was the relationship between designated nations and territory – the foundation stone for the creation of new states. If Yugoslavia no longer existed, and the rights of nations constituted the supreme principle, then Hungary or Albania could claim the right to annex territory where their nationals were resident. More seriously, the rights of Croatians to form a nation-state and the right of the Republic of Croatia to become an independent state were two very different propositions. To confuse them, and use the borders of the republics as a blue-print for new international borders in the name of *national* rights was both a gross confusion and highly irresponsible. If nationality was the supreme principle, then borders could be open to negotiation. Finally, a decision to solve Yugoslavia's problems by disaggregating it on national lines would destroy Bosnia, since Bosnian was not a nationality, though Serb and Croat were.

Many outside observers believed that there was a strong case for keeping Yugoslavia together in 1990–91, and that making available financial resources would reduce economic insecurity, increase the legitimacy of the federal government and allow the time for democratic organisations to emerge. If federal or confederal solutions failed, it would still take time to work out a new pattern of states and their borders, and find out who wanted to remain in a rump Yugoslavia. The EU was shaping policy on those lines, including the Badinter Commission which was working on the criteria (democratic rights, protection of minorities) which would be a precondition for international recognition of successor states. The important point was that there could not be a piecemeal solution, and nor should it be rushed. In the event this was precisely what happened.

At the Maastricht Treaty meetings of December 1991 the EU broke with the recommendations of its own commission and agreed to recognise Slovenia and Croatia as independent states. This was done under German pressure, and appears to have involved a complex deal, buying off British reservations with an opt-out on key aspects of the EU harmonisation policy (the social chapter), and rewarding Greek acquiescence with a veto on the recognition of Macedonia. In reality the Slovenian and Croatian governments had been receiving covert support for an independence strategy for a number of years – from Germany, Austria and the Vatican. Tudjman's party was receiving substantial aid from the Croatian diaspora in Australia and North America, many of them exiles from the Ustashe regime of 1945, keen to take up their historical mission where it had been broken off.

Recognition of Slovenia and Croatia – even secret support by a major external power – destroyed any federal solution to the crisis, and doomed Bosnia to a vicious war of dismemberment. The impact of external agencies did not finish there. At a later stage, as part of the Vance–Owen peace plan, maps were published delineating that dismemberment of the territory on 'ethnic' lines, accompanied by the impression that these might be revised – and this stimulated a further round of territorial war. Such maps show ethnicity in vivid colours: no other kind of belonging is visible in them, nor were representatives of non-nationalist political cultures invited to the talks. This does not mean that all the suffering experienced in Yugoslavia is the result of western intervention, or that all the interventions were pernicious (roughly the argument of Gowan 1999). Nor does it mean that international intervention followed a clear and consistent policy objective: the evidence from the early 1990s is of confusion, misunderstanding and inconsistency – though certain governments did know what they wanted to happen. Since 1995, and particularly in Kosovo, the situation may have changed, and international policy in ex-Yugoslavia may no longer essentially be concerned with what happens in the successor states; instead it has become a region where various post-Cold War issues are fought over and clarified: the relationship between the United States and Russia; the new role for NATO; the coherence and autonomy of the European Union (see Blackburn 1999). Those issues are being fought about over the heads of the Yugoslavs.

VIOLENCE AND ETHNICITY

The wars in Yugoslavia in the 1990s, for all their global importance, were in the end local wars. They were not fought between states (Bosnia against Serbia); nor were they conventional wars between armies, though the fire power of a conventional army was sometimes decisive in gaining or holding land. They were fought in the main by civilians, and by neighbours, in the villages and small towns of the Serb-populated areas in Croatia and Bosnia. The dominant groups in the post-communist governments sought to retain control over resources, territory and populations, proclaiming themselves the defenders of peoples who were (all) under threat. Given the continued existence of other traditions of solidarity in Yugoslavia, the ruling forces had to consolidate their power by dramatising a threat to the people that was immanent and devastating. Indeed they made it real, and the use of force was intrinsic to their political strategies.

As armed conflict grew, the character of the regimes and the interests they represented changed. The formal economy was devastated – during the war Bosnia's industrial output fell to 10 per cent of its pre-war level (Kaldor 1999: 49) – while swathes of previously important manufacturing and commercial activity disappeared. Their destruction and the resulting unemployment destroyed the revenue base, which in turn undermined the whole public sector. Millions left the country. In the end those who possessed weapons – and skills in the political use of violence – came to control the manufacturing base which survived (chiefly that producing armaments) and most forms of commerce. Young men were attracted to the militias for the spoils of war – not least to loot the destroyed houses. Those carrying arms controlled the trade in petrol, cigarettes, guns, drugs and alcohol. Military hardware itself, including tanks, was sometimes traded for deutchmarks between rival armies (Judah 1997: 250). Some of the fighting around Sarajevo was not for 'ethnic' gain, but for control of the strategic smuggling routes through which ran the immensely profitable black market supply lines into the city. Many of the larger towns developed a gangster economy, while in rural areas there remained farmers producing food, and armed forces. Some observers have borrowed the vocabulary of feudalism to describe such a society (peasants and warlords), though it can also be seen as a version of post-modernity.

The consolidation of these regimes in Serbia and Croatia involved two main stages. The first was a media campaign to convince the people of the immediate horrors that they faced unless they armed in defence. Television had been taken over early by the governments in the republics, and by 1991 ceased to have any pan-Yugoslavia coverage. It was used to replay the Second World War, showing scenes of Croat Ustashe attacking Serbs, Chetniks or communist partisans attacking Croats. Footage from the war was intercut with scenes from the present, including the disinterment of Second World War corpses from caves and death-pits. There was an overwhelming assumption about the historical continuity of the nations, and of the conflict itself – above all claiming that 1991 involved the same forces that had fought in 1941–45. The communist government had suppressed discussion of this period; now it all came into the light, and Yugoslavia filled up with the tombs of unknown warriors.

The second stage was the achievement of ethnically homogeneous states by moving people or borders. This was achieved partly through non-violent means – sacking state employees who were not from the dominant *narod* (Serbs in Croatia, Albanians in Kosovo), harassing and disenfranchising minorities (Hayden 1996). The second strategy was the use of terror, and when violence finally occurred it had been well prepared. Arms were sent to co-nationals in key areas and stockpiled. Militias were recruited – the most effective and notorious were Serbian, but they were also present in Croatia and Bosnia. The most famous was run by Arkan, a smuggler – reputedly also employed by the Yugoslav security forces to kill émigrés – and owner of the fan club of the Belgrade Red Star football team. He recruited fighters from the football club and the criminal underworld, and was supplied directly by the government with all his military needs.

There were various techniques for installing a reign of terror. In 1991–92, forces operating in Serbian villages in the Krajina would commit atrocities against Croats, provoke counter-attacks, and then bring in the JNA to take command of the district, thus bringing it under Serb control. The militias could also operate in the absence of the regular army, moving into a district with a mixed population, killing leaders and radicalising the population. In these contexts what mattered was not the abstract *narod*, but the dense local pattern of kinship and neigh-bourhood, which was forced to close ranks. A Croat, living in a village where a Croatian militia had killed Muslims or burnt them out, was forced into self-defence, and also into dependency on the militias. Cleansing destroys the social networks within which people meet; it is the extreme conclusion of the process of ethnicisation of social relations analysed in the previous chapter. 'The turning point in the dissolution of Yugoslavia as it affected endangered individuals, according to their own reports, was when they saw the necessity as families or localities to resort to guns in self-defence' (Woodward 1995: 391, 483). Fear is the key, and it was the militias who generated it, creating insecurity in order to sell security. Later the region filled up with people who, having lost their families and their homes, had nothing further to lose, and the dynamic changed. In some cases local conflicts pre-existed the war, but their escalation and devastation were the result of changes in power relations at a much higher level.

The process now called 'ethnic cleansing' occurred in areas with mixed populations, and this has led to a rather simple conclusion: that it is a manifestation of ethnic hatred (or that 'these people just don't want to live together'). However, virtually all published first-hand accounts indicate that terror was not ubiquitous in mixed areas, and that it happened for military reasons – where there were industrial assets or supply corridors. Cleansing was initiated as a means to achieve military ends, though once started other factors came into play. Secondly, some areas remained peaceful even when surrounded by fighting. In the absence of detailed ethnographic studies we have to rely on the evidence of perceptive journalists passing through. Misha Glenny remarks on the Serbian communities in eastern Croatia who came from 'old slavonian families who had no quarrel with the old Croatian families in the region' (Glenny 1996: 107–8). The contrast is made with the atmosphere in villages established after 1945 by transplanted Serbs from the Krajina and Croats from areas in western Herzegovina, which had seen bitter fighting in the 1940s.

After the militias or local armed extremists began killing, the use of violence usually spiralled. It included the rape of women – a frequent occurrence in any battle-zone, but one which took on a particular salience and horror in the Yugoslav conflicts. Many areas had patriarchal honour codes which had been reinforced by the collapse of state authority and the militarisation of the economy. Armed men guarded their own homes and those of the *nacija* neighbourhood, with its womenfolk. In this localised and highly personalised conflict, the killing of a man or the rape of a woman carried a parallel significance in terms of a group's ability to defend and sustain itself. More extreme abuses have been documented: camps where Muslim women were held and raped repeatedly, often until pregnant.

In the worst cases – and there were hundreds – the whole target population of a town or district were killed or driven out. Their homes were looted and burned; libraries, monuments, mosques or churches were destroyed so that no physical evidence remained of their existence. The most accurate verb I know for this process is the Italian *annientare*, 'to render into nothing'. A recurring practice (see Bax 1995) was the dynamiting of tombs and the flattening of graveyards: the classic 'places of memory' where individuals were connected to their ancestral lines, to collectivities and to territory. This was desecration – the destruction of what the enemy held to be sacred, essential to its being. It was also an attack on those objects which were central in the practices through which were constructed identities based on historical continuity. These symbolic connections had become apparent a few years earlier when, in 1989, the bones of Prince Lazar, the leader of the battle against the Turks at Kosovo field in 1389, were exhumed and paraded through the Serbian Republic: one of the undead who moved through Yugoslav society in this period.

CONCLUSION

Noel Malcolm, with controlled irony, reproduces the standard phrases used by western leaders to characterise the conflict in Yugoslavia. John Major told the House of Commons in 1993 that 'The biggest single element behind what has happened in Bosnia is the collapse of the Soviet Union and of the disciplines that that exerted over the ancient hatreds of old Yugoslavia' (Malcolm 1996: xx). The media reported that a civil war had broken out, fighting had flared up, while law and order had broken down. The declared objective of the Foreign Office was to do nothing that would 'prolong the fighting' or 'hinder the peace process'. 'The war was seen as essentially a military problem, caused by something called "violence", which "flared up" on "both sides"' (Malcolm 1996: 242). These influential representations of the war, with their extraordinary reifications, contain no political agents, no sense of the political purposes of the actors in the Yugoslav tragedy.

Historically Yugoslavia has been a military, religious and commercial cross-roads; a region of repeated population movements and a country of unusual cultural diversity. Some of this diversity came with the construction of states and empires within the geographical region, while some is associated with, and remembered in terms of, the historical continuity of distinct peoples. Nationalism is the political articulation of difference, not the difference itself, and over time the differences and the articulation may change. Some had believed that Yugoslavia itself as a state was defined by its plurality, or that plurality was a defining feature of being Bosnian. Indeed, a society where 83 per cent of the population spoke a close variant of the same language, and where the main lines of cultural division were religious, though the state itself was secular, had a good basis on which to consolidate a plural polity. In recent Yugoslav history it has been through war that nationalism has been consolidated.

By 1990 the accumulating failures of Yugoslav communism had come to a head. As a 'decentralised totalitarian regime' its economy had lost efficiency and gone into reverse; democratic reforms were blocked, and with them any chance of mobilising pan-Yugoslav opposition to the regime. Civic conceptions of citizenship were entering into crisis as the federation weakened and rights based on *narod* membership came to dominate in the new centres of power in the republics. Nationalism replaced communism as the dominant political and cultural movement – either as an offensive strategy to control people and territory, or as a defensive reaction to aggression. To be effective it had to mobilise existing cultural divides and everyday experience, and to tap into the world of kinship and neighbourhood.

In the political language of cultural nationalism in the 1980s and the electoral campaigns of 1990 ... the most commonly-used word politically, from Slovenia to Serbia, was *hearth*. The focal point of a home or homestead, hearth becomes a metaphor for property, community, citizenship and patriotism, all in one. (Woodward 1995: 237)

It is at the level of daily practices such as those associated with kinship that nationalism as a construction of identity, continuity and difference becomes powerful. But nationalism is not kinship, though it may involve the incorporation of kinship into a state-level discourse of race. The dynamics of this situation are best captured by Woodward (1995: 18): 'To explain the Yugoslav crisis as a result of ethnic hatred is to turn the story upside down and begin at its end.'

SOURCES

The preparation of this chapter involved sifting a much wider literature than for the other case studies, but I am aware that it is still very selective and that important new work is emerging all the time. There are radically different interpretations of the historical processes leading to the break-up of Yugoslavia. States are both normal and 'artificial': there is no state in Europe whose territorial boundaries and composition have not changed or been seriously challenged in the course of the twentieth century. There are conflicting views about whether Yugoslavia was 'viable' after 1989, given its internal dynamics; the chapter is written from the conviction (not shared by all) that there were more losers than winners from the manner of its disintegration, and that that will be true for a long time to come. For the international dimension of this process I would strongly recommend Woodward (1995); for nationalist dynamics, Brubaker (1996: 55–76); for the internal dynamics of Yugoslavia under communism, Dyker and Vejvoda (1996); for an overview of the Bosnian war, Kaldor (1999). Glenny (1996) is full of insight and information on everything from the villagers who refused to fight to the evolution of international policy. Silber and Little's *The Death of Yugoslavia* (1995) and the BBC television series of the same name document the descent into war with some extraordinary interviews with the protagonists.

The 'viability' of Yugoslavia is connected to a second controversial set of issues around the relative importance and stability of the national and 'ethnic' divisions within the country. It is easy to argue that because these divisions are central to contemporary politics, and conceived in deeply historical ways, they have always been present in the same places – the political bedrock under the surface. With the collapse of the state and local institutions, forms of collective action and solidarity which cut across 'ethnic' boundaries are remembered (if at all) as moments when people's true identities and interests were temporarily masked. The contrary view is that the boundaries are porous, shifting, present in some social contexts and absent in others; that they interact with other social divisions, and are dramatically reshaped by changes such as the decline of the Ottoman Empire, the end of feudalism or socialist economic policy. My own understanding of Yugoslav diversity has been informed by those who study the history of societies, not of nations (such as Dyker 1972; Malcolm 1996, 1998; Pinson 1996b); by work on peasant politics and the agrarian parties (Bideleux and Jeffries 1998; Jackson 1966; Mitrany 1951); and by the work of the anthropologists (Bowman 1994; Denich 1994a, 1994b) and others listed below. The disagreements about the importance of these boundaries are crucially also disagreements about the characteristics of ethnicity and nationalism, which is why the conceptual discussion was necessary.

Ethnographic research is essential to our understanding of these processes, but the variety of Yugoslavia compels specialisation. It would be possible to pursue the analysis through the ethnography of Serbia, from Halpern and Halpern (1972) through to Anzulovic (1998) and van de Port (1999), but I have concentrated on Bosnia because of its symbolic role in relation to the Yugoslav federation, and because everything was at stake there. There is less information on the involution of Yugoslav society in the 1980s, and it is striking how much that is now considered inevitable was not anticipated by even the most attentive and informed observers. Lockwood (1975) is very useful, partly because he was not primarily concerned with ethnic divisions. Sorabji (1989, 1993, 1995) and Bringa (1993, 1996) deal with the more recent period. The collection of essays in a special issue of *Anthropology of East Europe Review* (1993, vol. 11) is invaluable. It forms the basis for Halpern and Kideckel (2000), though there are significant additions, omissions (Olsen, Bringa) and changes in emphasis in some of the contributions, and this makes referencing difficult. Both collections show that bitter disagreements about what was happening can be found amongst professional anthropologists. One of the most harrowing and informative sources on the destruction itself is the Granada TV film, *We are all neighbours*, made with Tone Bringa.

8   OCCITANIA AND LOMBARDY: POPULISM RED AND WHITE

The last two chapters, on the Basque country and Yugoslavia, examined nationalist movements attempting to create new states through secession. There are many other kinds of nationalist politics, from the everyday, 'banal' (Billig 1995) articulation of national frames of reference in established states, to the struggle for greater autonomy by more inclusionary nationalist movements: Scotland and Catalonia are much-quoted examples. Two case studies cannot represent the whole, though they can serve as the basis for questioning certain kinds of generalisation – about the contrast between nationalist movements in western and eastern Europe, or the status of the contrast between 'civic' and 'ethnic' forms of nationalism. In the first part of this chapter I shall draw out some of the common political processes which have emerged from these two case studies, especially those leading to the affirmation of nationhood.

The remainder of the chapter explores two political movements which combine class and nationalist themes, one in the 'red south' of France and another in the 'white north' of Italy. Each of them articulates the distinctive qualities and rights of those who live in a particular place; at the same time each movement understands local societies to be stratified, and stratified in politically significant ways. They may describe this in the language of class, or as a more generic opposition between the people and the rulers. In some contexts this opposition is transposed into territorial terms, so that those of the locality or the region are 'the people' while the rulers are identified with the forces of the national economy or the state, external but encompassing. The politics of place, belonging and rootedness have a very long history in Europe, and have undoubtedly become more prominent in the last 20 years, partly in reaction to the dominant neo-liberal models of 'dislocation', labour market flexibility and mobility. They surfaced in many different movements, and although these two shorter case studies cannot document their full complexity, they can reveal the convergencies and divergencies between them, and the slippery slope between the rights of local people and the politics of exclusion and xenophobia. One way in which these more hybrid movements have been analysed is through the concept of populism, and the concluding section of the chapter will discuss the significance of this term.

THE MAKING OF NATIONS

Debates on the interpretation of nationalism have frequently pitched those who stress the historical antiquity of culturally distinct peoples against those who stress the construction of nations by political elites in a recent historical period. The discussion of political processes in the last two chapters has formulated the problem in a slightly different way, by focusing on the interaction between local and higher-level systems. Culturally, certain kinds of everyday experience, particularly those arising from the sphere of domestic and neighbourhood relations, have been encompassed and reformulated within a larger interpretative frame – that of a people with a distinct culture. Socially, certain lines of division, patterns of inclusion and exclusion intrinsic to the operation of the informal economy, to marriage patterns or religious observance, have been hardened and totalised into the boundaries of the nation – a process which in its most extreme form has become known as ethnic cleansing. These cultural and social processes are connected – inevitably so, since culture – in its widest sense (forms of knowledge, understanding and representation) is not free-floating, but embedded in social relations.

A concern with interactions runs through much recent work: between popular culture and the work of nationalist intellectuals; between 'ethnic groups' and modern nations. A. D. Smith has written on the need for interpretations of the building of nations which balance 'the influence of the ethnic past and the impact of nationalist activity' (1995: 16). In his critique of both those who see the nation as an unchanging essence and those who see it as a modern and ad-hoc invention, he draws attention to the role of nationalists in the 'rediscovery and the reinterpretation of the ethnic past and through it the regeneration of their national community' (1995: 3). There is much that I would agree with in his comments on how the historical narratives of nations emerge, and on their popular resonance, but overall the approach developed in these last two chapters has a different emphasis. We have seen the need to look critically at the connections between peoples and cultures, and the need to look at the power field within which national narratives are constructed. The material has also suggested not the regeneration of communities but the development of new social boundaries (including states), legitimated through a discourse which selects out a particular range of cultural markers and weaves them into a narrative of the unbroken history of a people.

In both the Basque country and Bosnia nationalism came to have predominance over all other political questions, though it did so in very different historical contexts and through very different forms of mobilisation. Nevertheless there are common themes, and I want to draw together some of the comments on the processes of social and cultural polarisation which have characterised political life in the two cases. In both accounts I gave details of local forms of sociality: the voluntary associations of co-operating neighbours (*auzoak*) and leisure-time activity (*caudrillas*) in the Basque country; the *nacija* of Bosnia. These were the more overt aspects of a general process of social inclusion and exclusion, operating in domestic, kinship and neighbourhood relations and in the informal economy. They built into a definition of who 'we' are ('us from here') and into a moral map of trust, intimacy and distance, versions of which were found very frequently in peasant Europe. They also incorporated experiences and relationships which became central to the narrative of nationhood, of the home and homeland – the framework within which this nationalist version of the good life was imagined – the hearth-fires, the warmth, solidarity and enduring relationships between peoples and places.

This is not, however, an argument for simple continuities in nation-building, nor for the primacy of 'primordial' ties: this is after all a very particular and romantic vision of the good life. These networks of rural producers were not the whole picture – we need to build in other dimensions. Rural households were part of a social world which included towns and markets; they had to wrestle with landlords and merchants, often people who spoke the same language and practised the same religion as themselves. They lived through periods of rural depression and industrial growth, which in the long run drew the majority of the population into urban centres and a monetised economy. For many such people, in many periods, rural lifestyles signified poverty, discomfort, backwardness and economic uncertainty; the future was the factory, a flat and a wage. These had never been homogenous societies, but new divisions of labour and different kinds of migration added other layers of social and cultural differentiation. But in both societies the 'motor' of industrial development was not a smooth machine: there was recession as well as growth, periods of high levels of unemployment and blocked social mobility. Sometimes people were able to maintain a foot in both the rural and urban world, or reactivate links into the informal economy, for all that there were renewed questions about what each world offered.

The nationalist movements we have been examining take hold when local social worlds become more heterogeneous, and older ways of gaining a livelihood enter into crisis – and with them existing forms of organisation, knowledge and values. Local forms of sociality are important both in themselves and as conceptual frames for 'nation-building'. However, if we refer to nations as 'kinship writ large', or talk of nation-building as 'the regeneration of community', we underestimate the transformations which occur. The movement appropriates some of the substance and the exclusionary mechanisms of kinship groups to create new groupings amongst populations within a much larger territory, amongst people who will never be kin or neighbours. It also creates solidarity amongst social strata where none had existed before, and ejects long-term neighbours from the moral community. We found all this most clearly in the discussion of Bosnia – in the nineteenth-century alliance of landlords and peasants in a Muslim political mobilisation, in the complex history of 'Muslim' as a state *narod* category, and in the manifold differences between state and local categorisations. We also saw how political polarisation in the 1990s turned long-term neighbours with their overlapping cultural repertoires into accidentally co-resident representatives of global religious communities.

Political mobilisation has normally come in the first instance from outside the rural networks, but it has to capture each locality within the territory of the nation and speak for the majority, or the 'true people', who live there. In the Basque country this came about through the creation of clubs and political parties linked downwards into local informal associations, and federated horizontally to dominate the political life of the provinces. Cultural revivals, publishing, festivities, schools, and the activism of clergy all helped to create the linkages and the dense network of parallel organisations within which the patriot lived. Similar movements were found throughout Europe, including Yugoslavia in an earlier period, but in the 1990s it was the militias and military leaders who precipitated the faultlines of the nation.

Nationalism famously naturalises its own existence, and in the kinds of movement we have been examining this is accompanied by moves to downgrade and destroy forms of solidarity which cut across its borders. This occurs through attacks on the work of rival political movements, through social ostracism and, in extreme cases, through the strategic use of violence against people and property, forcing local populations to take sides. We have seen the end result of all these processes in the segregated Basque village, where education, worship, kinship and the economy were organised in parallel systems – one for the patriots, the other for the despised immigrants and the unpatriotic supporters of alternative political visions. We have seen it even more destructively in Bosnia, where 'ethnic' polarisation has driven populations into separate territories and destroyed the contexts in which they once used to meet and work together: the factories, co-operatives and universities.

Nationalist movements create new forms of solidarity and new lines of social division, but what happens in the cultural field as this polarisation takes hold? We have seen the shifting definitions of Basque culture, and the multiple, overlapping cultural repertoires found in Bosnia. The nationalist identity narrative must produce, instead, a set of representations which portray a people who share a common way of life and historical experiences – a culture which is set apart, in its essentials, from those of other peoples. One commonly found response is to locate these essentials in the peasantry – not just because of the political importance of rural society in the relevant periods, but because they could be seen to carry more of the important qualities (purity of lifestyle, continuity of practice) than townsfolk. But portraying rural peoples as carriers of the authentic culture of the nation is not without problems. To work, it has to render invisible both those aspects of culture which Basque or Croat peasants share with other peasants, and those cultural repertoires which belong to peasants but not to bankers or teachers.

This means that certain discursive strategies are necessary in delimiting the culture of the nation. The word 'strategies' may evoke a conscious and manipulative process, and in fact there is often experimentation in the work of nationalist intellectuals, and quite sharp shifts in the definition of the nation. However, overall what should be stressed is that certain themes emerge over time as more resonant, and more effective, in drawing the boundaries in the right place. 'Home' and 'homeland' unite a wide social spectrum of those who consider themselves to belong in a place, against 'others' who are conceived primarily as migrant and rootless; more than this, they make irrelevant the fact that members of the two groups are actually working alongside each other in the same factory. Very often these movements give prominence to one specific dimension of a nation's culture which stands for its identity, and can stand out as a marker of difference. These dimensions include the compressed symbols and shrines which evoke a range of meanings and which, through their continued public and private use, evoke also the memory of specific experiences; memories ordered and lit by the history of the nation. As a discursive strategy these compressed symbols are making the part stand for the whole (in rhetoric this is called synechdoche). The part is usually less problematic than the whole, and this is one of the ways unity becomes visible and diversity invisible.

Movements may have to create unity within populations which have very few horizontal social linkages (in rural districts); or who speak mutually unintelligible versions of a language (as we saw amongst Euskera-speakers); or who practise significantly different versions of Islam (as we saw in rural and urban Bosnia). In this context I have referred to a discursive strategy based on homology: the articulation from a morally central space of an opposition between 'us' and 'them' in cultural terms, which people could recognise in their various local realities. Crucially, what people share, over and above their differences, is a state of oppression. This is only the beginning. The work of the movement itself creates not just new forms of solidarity – the social networks within which the nation exists – but a growing literature and a repertoire of cultural forms within which it is expressed. If the movement culminates in the formation of a new state with its own education system, media, and power to standardise language, then it has achieved control over the commanding heights of many cultural domains.

Commentary on definitions of the nation's culture and their subtleties should not obscure the fact that in many circumstances the nation is defined as a race. Racial discourse is more powerful and dynamic than those discourses built around the fate of minority languages or cultural traits. It leads straight to notions of purity, continuity and the ways these are threatened; it creates a gendered narrative with historic missions for men and women; it stands for unbridgeable difference between nations, since culture can be acquired but descent lines cannot. Race can stand alone, but it also underwrites many of the cultural definitions discussed so far. Roger Just (1989: 76–7), writing of Greece and the concept of *ethnos*, summarises very well a discursive strategy which is frequently found in the definition of national identity: all the arguments about historical continuity, occupation of territory, language and culture are important and fought over, but they constitute *evidence* of belonging to an ethnic group, whereas the group itself is *defined* elsewhere – in blood, descent and race. Within this frame, the absence of historical evidence does not discredit the national narrative, and the absence of a particular cultural 'marker' does not necessarily lead to a person's exclusion from the national body, because that evidence and that membership have been established elsewhere. Some of the characteristics of nationalist arguments about identity – the ability to shift ground on cultural issues, the presence of elliptical references and tacit understandings – become more comprehensible in this light.

It is not easy to assess how important racial discourses are in the overall construction of nationality, and how this varies between cases and over time. We could say, somewhat provocatively, that in the late nineteenth century nationalists talked so much about race because the rhetoric of shared culture was implausible in societies strongly marked by a rural/urban divide; from the mid-twentieth century onwards nationalists talk so much about culture because in many circumstances they are not allowed to talk about race. Substantial differences do exist within European nationalisms, but it may be better to move from simple typologies (for example, the dichotomy between civic and ethnic versions) towards a view that there are many strands in nationalist discourse, but that they are present in different ways and their relative weight varies between societies and over time. A society may be committed officially and legally to a civic and inclusive approach to nationality, but contain within it a strong covert and commonsensical view of belonging based on descent, in defining who is 'really' British or French. In certain periods, associated particularly with the politicisation of migration, this breaks through into mainstream political culture, through attempts to change laws on citizenship, and through the work of populist political leaders keen to stress once again the existence of unbridgeable difference.

Populism and the renewal of nationalism within western Europe are themes which also emerge in the second part of this chapter. It is made up of two shorter examples of recent contemporary movements which combine a series of economic demands with appeals to the cultural identity of those who live in a territory or region. They are thus hybrid in terms of the normal typologies, but it is my hope that the analytical work of the earlier chapters will make it easier to interpret them.

RED SOUTH

Occitania is a territory defined, in cultural terms, as those lands in southern France where versions of the *langue d'oc* are spoken. It was never politically unified into a state, but it was a region with autonomous political and cultural forms which came under French domination, most notoriously after the crusade against the Cathar heresy in 1209. These massacres were precisely amongst those events which, Renan argued, had to be forgotten if the French nation was to exist. The Occitan activists have chosen to remember them. Occitania has fuzzy boundaries and no standardised culture; in practice it is conceptually unified around an opposition to the French language and the French state. The foundation moment in Occitan identity narratives does not establish the origins of the people, but the moment when 'they' were subordinated to the French. The cultural, and not least linguistic, diversity of the Occitan region need not stand in the way of a unitary identity, if this is based on a common insubordination to France; on homology, as was suggested above. That is, each local population recognises itself in a narrative which opposes 'our' way of doing and being to that of the dominant French.

In the late nineteenth century there was a literary renaissance of the *langue d'oc*, under the leadership of the writer Mistral (who won the Nobel Prize in 1904). The first cultural movement, the *Felibrige*, which gathered pace in the years following the establishment of compulsory eduction in French (1885), celebrated the language and cultural values of an eternal and mystical Occitan land (see Grillo 1989: Chapter 4). One hundred years later one strand in the Occitan movement continues to celebrate local rural traditions through festivities and a range of cultural activities. Metropolitan intellectuals often castigate these activities (and the young who settle in these rural areas) as nostalgic, inauthentic, or worse. The strong antipathy to these reinventions is itself curious – it would seem that only interest in the wrong kind of culture counts as nostalgia. However, these activities do contain some paradoxes. In the world of MacDonald's, enthusiasm for the rural and the local, or what the French call *terroir et tradition*, probably does not indicate a desire to live within the bounds of the local, doomed to eat lentil soup in perpetuity, but to range more widely as a post-modern or translocal gourmet. Touraine also points out that some key symbols of this Occitan civilisation 'of pastis, rugby and bulls' (1981: 168) are decidedly masculine, and that in general rural idylls are less alluring to women.

Although Mistral's movement had been in favour of federalism, it had been politically divided between right- and left-wing currents, and little concerned with economic programmes. Most of the action groups which emerged after 1968 were mobilised primarily around the economic problems of Occitania: unemployment, depopulation, an ageing population, low levels of industrialisation. A gulf was opening up between this region and the heartlands of the French economy. Touraine believed that

France has renounced its social, economic and cultural integration, and has accepted ... dualisation. The Midi, or south of France, is becoming a 'Mezzogiorno', a controlled drift towards underdevelopment, occasionally disguised behind deceptive labels such as the opening up of tourism, the sweet life, or even defence of the patrimony. (Touraine 1985: 164)

Emigration was for many the only alternative to unemployment at home, and one of the most important political groupings was significantly titled *Volem vivre al pais* (VVAP): 'We want to live (and work and decide) in our own country.'

Responsibility for this economic crisis was attributed to the French state, which had variously failed to protect the interests of the region or guarantee the same rights to all its citizens, and presided over an economic system which had drained the territory of labour and productive capital. The French state, when it was active, was seen as an invading force, riding roughshod over local interests; in fact the most sustained and violent protest centred on the extension of a military base at Larzac. The solution to the current ills of the region would involve transforming the relationship with the highly centralised French state, whether this took the form of a change in planning policy, federalism or autonomy.

On the economic front the most important actors were the wine-growers of Languedoc, and we need to look in a little more detail at their history, because it is amongst this group that a class discourse emerges within the general demands of the 'Occitan people'. The integration of the Midi into national and international markets in the nineteenth century destroyed much of the textile industry and many forms of agriculture, but the growing railway network allowed the lower Languedoc to develop an intensive vine monoculture between 1855 and 1870 (Kielstra 1985: 252). There were some large domains worked with wage-labour, but most vineyards were owned and worked by family farmers, while in the first half of the twentieth century a network of co-operative *caves* developed to handle wine production. There were decades of prosperity, and then of crisis, resulting from the spread of phylloxera and later from over-production. Small producers have looked to the French state to guarantee their incomes by intervention in the market (distilling surpluses) and by protectionism against outside competition, most recently from within the European Union. Kielstra describes the region as a 'relictual space'. Smallholding farmers survived the first impact of the market, creating an intensive agriculture and maintaining relatively large and socially heterogeneous villages (Kielstra 1985: 258), but the region does not have the resources or opportunities for further development. Like certain areas of Italy's Mezzogiorno, households are forced to generate a livelihood from a variety of sources, and are heavily dependent on transfer funds from the state.

Since the late nineteenth century the labourers and smallholders of the Languedoc have generated a radical left political culture. The Socialist Party was strong but, like Puglia and Andalusia, in the early twentieth century this was also a land of revolutionary syndicalism, rejecting representative parliamentary democracy in favour of general strikes and tax strikes, mobilising a wide spectrum of the population. Eighty years later the economic demands of the small wine-growers were still central to the political agenda, along with a tradition of direct action. Both aspects emerge in Winnie Lem's (1994; 1999) analysis of the *Midi Rouge* and class politics. Men and women work long hours in the vineyards for a low income, and supplement it with wage-labour, both before and after they inherit the holding. Even when self-employed, they consider themselves part of the 'minimum-wage' category of the French economy, and have built 'a tradition of self-positioning as members of the working class' (Lem 1994: 410). This ascription is oppositional – intrinsic to it is the conflict between those who produce and those who do not. The term *exploitation* dominates their political vocabulary: as producers they exploit the land (*exploitation agricole*); as a class they are themselves exploited. As farmers they are rooted, and their language and practice 'resonate with the associations of locality, local history and local understandings' (Lem 1994: 403).

Smallholders attribute their low incomes very directly to the failures of the French state and, while workers go on strike, farmers demonstrate and protest. This tradition of direct action includes the blocking of transport systems, raiding supermarkets, and daubing state buildings with graffiti in Occitan. Lem argues that for this section of the population – and in this period, with the French left in power – the referents for class and for (local) culture were interconnected, transforming and reinforcing each other in political action and in everyday life. 'Occitania is more than merely a cultural construct; it is thoroughly invested with political significance' (Lem 1994: 405). For other categories of the population, and at other times, class and local culture diverge.

Touraine and his colleagues have devoted a series of studies to the difficulties faced by political movements which attempt to combine class and ethnicity, as we saw in Wieviorka's account of Basque nationalism. He believes that the Occitan movement is inevitably fragmented. One strand is concerned with affirming the rights of the Occitan people and saving its culture, although its efforts have been forward-looking, and the nostalgic elements which dominate other national movements are here more marginal. The second strand is more concerned with regional and economic underdevelopment, and seeks solutions through transformations within mainstream French politics. The movement is able to combine the different orientations but not unify them, though 'the Occitanist movement becomes reinforced each time it draws closer to the basic communities, at the elementary level where the French Left and Occitanan culture merge in order to oppose the centralising, capitalist, and bureaucratic state' (Touraine 1985: 161). This is the revolt of the '*pays*' against the neo-liberal state, and the unity of the themes within local experience is confirmed by Lem's ethnography.

The word 'populist' is used constantly to describe this kind of movement, and Touraine devotes long sections of his study to this theme. Populism is a loose concept in political science, and we can return to it after looking at a second 'hybrid' movement, one from Italy which reveals a similar regional mobilisation against the central state, but with a very different coloration and programme.

WHITE NORTH

The *Lega Nord* erupted onto the Italian political scene in the early 1990s and won sufficient support in 1994 to form part of a government coalition with other parties of the centre-right (Berlusconi's *Forza Italia* and Fini's *Alleanza Nazionale*). Only eight months later the leader, Umberto Bossi, broke the coalition and marched his followers back into opposition. This was a period of great turmoil in Italian politics. The collapse of communism in the east spurred a strategy of renewal within the Italian Communist Party which divided its supporters, while the end of the Cold War deprived the Christian Democrat Party of one of its reasons for existence. The convergence criteria for the formation of a single (European) currency, established at Maastricht, created pressure for fiscal reforms which would reduce tax evasion and cut public expenditure, while there was also pressure to eliminate forms of economic protectionism. Finally (and these processes are almost certainly connected) a series of corruption scandals ('*tangentopoli*') brought down much of Italy's business and political elite and wiped out most of the smaller parties. A political system based on stable and long-term allegiances evaporated in a few months, and one of the principle winners was a maverick party which dared to suggest that the formation of Italy, the sacred Risorgimento, had been a terrible mistake.

In northern Italy from the late 1970s onwards small groups of scholars established associations – *leghe* – which celebrated local dialects and culture, advocating federalism and opposition to the policies of the centralised state (for a good early history see De Luna 1994). Bossi was active in the Lombard League, mobilising young people in a protest movement which fought local elections and devoted a good deal of energy to establishing the existence of a Lombard people. At one level this was a difficult and surprising enterprise: Lombardy did not have the stable political boundaries, linguistic homogeneity or level of regional pride found in Tuscany, for example. Some progress was made, however. Flags, symbols and claims for a Lombard language emerged, while the rally of Lombard Nobles against the Holy Roman Emperor Barbarossa in the twelfth century is now celebrated with much pageantry every year at Pontida. However, this more folkloristic emphasis on Lombard ethnicity was combined, as we shall see, with arguments about the economic and political rights of the region. It became clear that it was possible to mobilise around a strong sense of localism in the villages and small towns without getting into dubious arguments about the common culture and history of Lombardy. People would revalorise their dialects, both as a means to social inclusion, and in acts of resistance against the Italian state. They could also revalorise community, locality, local knowledge and practice against a state machine which was too remote to see what it was doing. Localism, and an oppositional stance, were at this level more potent than claiming one language or culture. The structure of the movement – a federation of leagues in a territorial hierarchy – reflects this.

Bossi, who dominated the Lombard League and was elected Senator in 1987, welded all the leagues from Piedmont to the Veneto into the *Lega Nord* in 1989, and this became a major player in national and regional politics. We can now turn to the social composition of the movement and its political programme.

In a series of influential studies starting in 1977, Bagnasco has argued that Italy is internally divided into three territories: an underdeveloped south; the 'Fordist' industrial triangle of the north-west; and a third Italy in the centre and north-east. This third Italy is a territory of prosperous agriculture and, since the 1960s, of dynamic small-scale industry, which has created a series of specialised industrial districts in an urbanised countryside. Although the once important division between town and country is much eroded by this pattern of industrialisation, there remains a tradition of localism, of co-operative networks and local savings banks. There is a pattern both of flexible production systems and of social mobility, between waged employment, self-employment and entrepreneurship. Although Bagnasco stressed some common economic and social characteristics of the 'Third Italy', there were important differences. Rural Lombardy and the Veneto had been dominated by peasant smallholding and a 'white' Catholic political culture, Emilia-Romagna and Tuscany by the share-cropping system and 'red' communism. The *Lega Nord* took root predominantly in the white areas, without a history of class mobilisation, after the collapse of the Christian Democrat Party.

Electoral support for the *Lega Nord* waxed and waned in the 1990s. During its high points support came from across northern Italy, including the big cities like Milan, and also from a very wide social spectrum. During the low points the *Lega*'s electoral base was concentrated in the mountain villages, small towns and urbanised countryside of the Veneto and Lombard provinces, becoming 'the party of the northern industrial periphery' (Bagnasco and Oberti 1998: 161). The activists of the *Lega Nord*, as in Occitania, are predominantly young men who use sexualised political slogans, and engage in street-level contestation, painting graffiti, changing road signs, heckling opponents. Whether wage-labourers or tied into family businesses, in this small-town environment they tend to know each other as friends or schoolmates and socialise together in the bars, pizza parlours, pin-ball arcades and sports centres of this hinterland. Much of the normal apparatus of internal party democracy – committees and elected officers – may be missing or ignored, but analysts have noted the dense network of meeting places and activity which characterise the *Lega* as a movement, and its habits of direct participation and unmediated, populist political language.

The *Lega* is notoriously volatile in its proposals. Political scientists try to make sense of these changes by talking of phases in the movement, but the evidence is not always convincing. For example, the stress on ethnicity and the racist attacks on migrants are said to belong to an early, immature stage of the movement, before it embraced a more realistic programme. Unfortunately the *Lega* continues to legitimate racist practice both in the speeches of its parliamentary deputies and in its grass-roots activity. Rather than an analysis of phases, I would argue that the *Lega* has cumulatively adopted and developed a series of 'common-sense' understandings about locality, migration, politicians, taxes and the state, and fused them into one self-confirming and self-reinforcing discourse. It has particular resonance in small centres with a tradition of civic pride, entrepreneurship and a strong work ethic, which have been through very rapid post-war industrialisation but now face growing economic insecurity.

The problems these regions faced in the 1990s came from two directions. One derived from changes in the organisation of production and distribution. Benetton no longer produces many of its garments in the Veneto; supermarkets came late to Italy, but again they do not buy locally, and are decimating the small retailers. Capitalism does not arrive in one hit, but impacts on different sectors in different periods, and although there will be winners as well as losers in the widening of markets and horizons, change is very fast and requires collective resources to handle it successfully. The second problem for the small producers and the self-employed comes from fiscal policy. In the Italian private sector, tax evasion is a national sport, but became more difficult in the early 1990s, after Maastricht. Computerised cash registers, controls on VAT invoices, income tax calculated on estimates of likely earnings rather than self-declaration, and a variety of other measures all had their effect. Stricter controls on the hiring of labour – employers' contributions double the wage bill and were widely avoided – squeezed profits in small businesses. The new tax regime was labyrinthine, time-consuming, sometimes punitive, and very unpopular.

If tax revenues are wasted they become even more unpopular. A second common-sense theme the *Lega* articulates is that the Italian state is inefficient and corrupt, financing an over-inflated, unproductive public sector which delivers very poor services. The revolt against bureaucracy is combined with a very specific attack on the way the hard-working productive north subsidises a parasitic south. The centre of this perverse mechanism is Rome, which becomes a complex symbol of all that is wrong in contemporary Italy. Rome is the centre of the existing political system, in which the parties are alienated from the people – a system which took control of the resources of the state to buy votes through clientelism in the south, and was exposed as universally corrupt by the scandals and trials of 1991–94. It is all a *mafia*. Rome is also the sink into which northern wealth is drained. In general there is little sympathy for using state revenues to redistribute wealth, and this is combined with a vehement attack on the moral qualities of southerners (*terroni*) and southern society. The attack is widened to all migrants, who are portrayed as parasitic – recipients of welfare, beneficiaries of citizenship rights which local people are losing, alien people who should live at home amongst their own kind.

The *Lega* works to construct a political, cultural and moral gulf between the north and the south of Italy, and in that sense to undo the Risorgimento and the commemorative pieties of state rhetoric. Part of that work involves patrolling boundaries, beating the bounds (organising a flotilla of boats down the Po river), boycotting institutions like state television, and also controlling the territory against the enemy within – most recently Albanian migrants who are accused of destroying the fabric of society. There is also a conceptual repositioning in *Lega* discourse: 'The further we are from Rome, the closer to Europe' runs another slogan, while *La Padania* newspaper is subtitled *Nord Mitteleuropeo*. If Italy was formed by an ill-considered yoking together of societies with very different cultures and values, the strengthening of the European Union (although often contested), and the erosion of nation-state sovereignty allow some of the damage to be undone. The European Union itself has generated a powerful (rather nineteenth-century) discourse of modernity to describe its own dynamic, and this provides a pervasive framework for describing social and cultural difference both within the Union and at its boundaries: difference is conceptualised within a hierarchy of modernity and backwardness. Thus northern Italy (like Slovenia in the previous chapter) is a society of hard-working, prosperous and progressive people who can draw 'closer to Europe' by freeing themselves from the shackles of a backward, parasitic Mediterranean south.

The *Lega* has two political options: separatism and federalism. With the first the *Lega* becomes a nationalist movement, stressing ethnic difference, and launching a campaign of civil disobedience and the kind of political theatre (issuing green uniforms and Lombard currency) which indicate a demand for independence. With the second the *Lega* allies itself with other political parties in the Italian mainstream to seek constitutional changes which would lead to greater autonomy, and plays down some of the ethnic themes. The second strategy requires a stable coalition with both *Alleanza Nazionale* (whose commitment to a strong state and its southern electorate cannot be finessed indefinitely), and *Forza Italia*, which competes directly with the *Lega* on a neo-liberal programme.

The *Lega*'s claims are often ridiculed in the rest of Italian society. Its leader, Bossi, has championed an unsophisticated (*rozzo*) political style, and the party is written off by political commentators with every slump in electoral support. It bounced back in the regional elections of 2000, declined in the national elections of 2001, but did enough to claim important ministerial positions in Berlusconi's second government. (Political science writings on the *Lega* include Piccone 1991; Ruzza and Schmidtke 1991 and 1993; Visentini 1993.) One aspect of the *Lega* which is difficult to assess is the switching between the politics of independence and that of reform: is that indecision, inconsistency, or the gradual emergence of a more realistic political programme? It may be none of these things. It can be argued that, as in the Occitan case, the movement combines two themes, one seeking to take control of the region by reaffirming the social values and networks of local society, the other championing a neo-liberal programme of tax cutting and market freedoms within Italian society. The balance between the two is unstable. If the *Lega* moves too much into the mainstream it loses touch with its young anti-state activists; if it gives too much space to folkloric localism and separatism it loses ground to Berlusconi's version of neo-liberalism, which is already more representative of the major business interests in the north. So support fluctuates wildly, but in its heartlands these two themes can be combined, and even reinforce each other in the *Lega*'s discourse: northern people are industrious and independent, they pay excessive levels of tax to subsidise those who will not work; this regime of a 'nanny-state' (*assistenzialismo*), whose benefits flow to the south and to migrants, is the creation of a corrupt political elite, who are allied to a corrupt business and financial elite; the state is external and disruptive, it is counterposed to a bounded local society, of those who know each other, know their own business and their territory, but who perceive, in a period of rapid change, that they no longer control their own destinies. The unrooted, disorderly and invasive migrant also becomes a focus for the construction and regeneration of territorial identities (for example, through the formation of local action committees), in a pattern which is becoming familiar throughout Italy.

POPULISM

The *Lega* has been described as populist, or neo-populist. There is some ambivalence amongst political scientists about the usefulness of a term which has been used to describe phenomena as diverse as the agrarian parties of eastern Europe or North America in the early twentieth century, as well as the regional movements and xenophobic parties of contemporary Europe (Taggart 1995). 'Populism' evokes a broad-based and radical movement of 'the people', radical in the sense of mobilised in opposition to existing power structures. Who are 'the people' in populism? If liberalism addresses the citizens, and socialism addresses the workers, populism represents hard-working, independent producers – 'the little men' running farms and family businesses who form the healthy bedrock of society. They are struggling against big business and finance (often portrayed as an alien and conspiratorial community), or against the state, which neglects their interests, taxes them without any return, and fetters them in regulations. We can think of populism as a type of politics, but also as a political discourse which can be combined with a number of political projects, and it is this range of combinations with makes populism so pervasive and ill-defined. It can combine with nationalism, and with versions of socialism, as we saw in the Occitan movement. It can also combine with neo-liberalism, as in British Thatcherism or on the Italian right, where policies designed to get 'the state' off our backs or rid society of welfare scroungers are conjoined with notions of organic communities, and a strand of anti-modernism.

Populism is said to emerge when groups are alienated from existing mainstream parties, and have come to distrust representational politics, since their representatives end up getting co-opted into the establishment and betraying their roots. There is a move towards more participatory forms of democracy, for leaders who maintain an unpretentious lifestyle, use vernacular speech and maintain 'unmediated' contact with their people. Bossi, for example, is a tireless speaker at small-town meetings and larger rallies. The moral dimension of populist discourse is polarised around an opposition between the honest, hard-working people, with a rooted and authentic culture, against the corrupt, sophisticated, cosmopolitan elites, a theme which has sometimes been articulated around an opposition between the country and the city, and sometimes around anti-semitism. An enthusiasm for economic liberties may be combined with social intolerance and political authoritarianism.

Two other aspects of these movements are particularly interesting in relation to the major themes of this study. If class movements have stressed identities constructed through work, and nationalism had stressed the shared homeland, these movements combine them. Both the Languedoc wine growers and the *Lega* activists point to their hard work and their role as producers as features which set them apart from others, but they also demonstrate a strong attachment to a land and its culture. Their demands are best summarised as *they wish to be masters in their own house*, and the movements have indeed produced songs to that effect (Baier 1991: 83). Moreover, as Narotzky (1997: 198) notes, 'The concept of "*casa*" is one that pervades the political construction of a Catalan national identity', though she goes on to disentangle the complex relations compressed into that homely symbol. The combination (found also in the populist peasant parties of eastern Europe) becomes comprehensible when we remember that the movements have at their centre, sometimes more in their imagery than in reality, people whose economic activity is based around the household, involving continuity over generations and within a territory.

The second point is that although these movements uphold the rights of self-reliant small producers, they emerge in a period of changing economic horizons and growing uncertainty. Two authors have addressed this issue. Touraine has argued that populism emerges as a desire for continuity in a period of change – sometimes defending a collective experience which is threatened, sometimes trying to control the process, but always failing to come to terms with historical ruptures (Touraine 1981: 172). This makes the identity narratives quite complex. They are similar to some of those found in unambiguously nationalist movements, with their foundation myths in the battles of the twelfth or thirteenth century, and a minority of supporters drawn to the resurrection of organic rural communities. But the majority of supporters in the Occitan and north Italian movements do not believe that the good society lies in the past and, in general, populism *per se* pays less attention to history than nationalism does – either to establish identity or as a basis for its claims. These movements are more oriented to the present, or at best to an ameliorative (rather than transformative) future, attempting to achieve growth, progress and security within the parameters of existing economic forms.

Populism is also a theme in *Integral Europe* (Holmes 2000). The most fascinating and challenging parts of this study analyse the way in which a whole stream of contemporary movements draws on what (following Isaiah Berlin) Holmes calls a 'counter-enlightenment' European intellectual tradition. He documents the way in which leaders such as Le Pen in France attack the notion that progress is the secret of happiness, advocating instead the pursuit of order and harmony (2000: 36), casting himself as a cultural healer and claiming authenticity through an appeal to experience and instinct, not reason. While the analysis of this political tradition is brilliant, Holmes says rather less about the movements themselves and their social base. However, he does trace the rise of integralism and populism to a number of general processes, in a way which parallels the comments of Touraine. He refers to two in particular. 'Fast-capitalism' involves rapid technological change which, combined with privatisation and deregulation, reshapes the landscape of wealth and deprivation with astonishing speed. At the same time it impoverishes many of the existing frameworks of social solidarity, and even the notion of society itself. Secondly, the European Union's drive to economic integration has led to the inadvertent creation of cultural pluralism which, together with the dynamics of capitalism, has undermined sovereignty at the level of the nation-state.

There is a growing range of movements arguing for the rights of specific peoples in all parts of Europe. Some articulate ethno-nationalism, combined with very strong opposition to mass transnational migration, like those in Austria, Belgium or Switzerland. For these xenophobic movements there is evidence of increasing pan-European co-ordination and, as Holmes (2000: 199) suggests, the path has been cleared for their entry into the mainstream of European politics. They exist alongside movements articulating claims for greater regional autonomy around more inclusive or civic forms of nationalism. Overall, these movements vary in terms of their social composition, identities and political actions, but they also display a series of overlapping themes and family resemblances. We can analyse similarities and differences to produce various political categories, but some of the analytical problems arise from precisely the combination of similarities and differences. One example is the emergence of arguments about the rights of those who belong in a locality. VVAP ('We want to live and work and decide in our own *pays*') defended those who wanted to remain in places which have become impoverished and marginalised. As a movement it embraced some of the ambitions of rural anarchism, but discourses about belonging are found across the political spectrum, ending with Le Pen's 'France for the French' and equivalent slogans in the *Lega Nord*. We can find movements for local and regional autonomy which start with open and democratic purposes, but develop a harder edge and a more exclusionary view about who belongs in the region. In the concluding comments I will take up some of the problems in categorising these recent developments.

The *Lega Nord* represents the wealthiest region of Italy, and one of the most prosperous in Europe – its average per capita income is close to that of Germany. Generally it embraces economic liberalism and social closure: it does not wish any of that wealth to be redistributed by the state, either to Italy's poorer 'peoples' or in welfare provisions for migrants. If redistribution policies within Italy are unacceptable, we can expect little support for redistribution at a European level. There is little interest in the historical connection between northern wealth and southern poverty, and the mechanisms through which they were generated in the period since the Risorgimento: instead they are attributed to accidents of geography and mentality. Occitania is one of the poorer regions of France – more so in the heyday of the movement 25 years ago than now. Many of the activists in the movement are also small family producers, but with very low incomes and locked into a declining sector of production. They mobilise against the French state, but in favour of greater intervention. In northern Italy the *Lega*'s policies merge with those of the major industrial and financial companies: de-regulation, privatisation, cutting labour costs and welfare programmes. In Occitania there was more convergence with the policies of the traditional left, in favour of redistri-bution and state investment, though we have to note that demands for protectionism (against cheap wine imports) come at the expense of even more disadvantaged parts of the world.

Localism and regionalism may be re-valued in periods when hopes for stable employment are frustrated, and inherited businesses and farms become valueless. To 'live, work and decide in one's own country' is a fine ambition, even if best stripped of illusions that the *pays* will be a static or harmonious community. Whereas Touraine has denounced the populism of these movements, on the basis of a rather purist conception of what constitutes class politics, Lem has insisted that Occitan regionalism (with others) has incorporated a strong class identity (Lem 1999: Chapter 8). Narotzky (1997: 220) has also suggested that territorial entities with a 'counter-hegemonic local culture' could be the basis for new class projects, through which people attempt to 'own their futures' (1997: 218). Her comments emerge from analysis of a district in Catalonia which contains farming households with multiple incomes, as well as textile production organised through sub-contracting, outwork and co-operatives. This is not a wage-labour economy, but people do have a practical consciousness of the multiple and interconnected ways in which their lives are shaped by capitalist relations. If class mobilisation based on wage-labour and national unions and parties has declined, perhaps a new movement can be built out of groups of people who are subject to capital in more heterogeneous ways, but share strong territorial links and an anti-state counter-culture.

Here I return to the combination of similarities and differences in these cases. Movements which articulate strong local identities, defining themselves in opposition to a centralising nation-state and which mobilise the 'small business' sector and households which often have multiple sources of livelihood, can nevertheless be committed to very different political objectives. We need more detailed research and analysis to illuminate why that should be. The 'small business' sector is an increasingly complex reality: while entrepreneurs and the self-employed may value independence in the workplace, they usually have very little control over the larger field of production and the markets in which they operate, the clothing giants for which they are subcontractors, or the wine wholesalers they supply. Small businesses are integrated into this larger economy in different ways, and they can interpret that integration in different ways. Increased market integration can be seen as the creation of a competitive and open environment, an opportunity for growth, or as working to the perpetual disadvantage of local people, undermining the autonomy they possessed when markets were less developed. Much will depend on the sector of the economy these small businesses occupy, but these interpretations will also be shaped by pre-existing Catholic or socialist political cultures, with their discourses of identity, solidarity and social justice.

Talk of regions is pervasive in contemporary European political discourse, and within the European Commission they are seen as a way of developing a legitimate level of government below the nation-state (Le Galès and Lequesne 1998: vii). Regions have the same warm glow and conceptual fuzziness as did 'communities' in the 1980s. The regions which appear on maps of Europe and in the minds of political activists rarely correspond to significant economic units, though this is not to deny that certain regions are taking political and institutional shape, and are buttressed by arguments about cultural diversity. The political question is what happens to 'populism' as Catalonia or northern Italy gain greater political autonomy (control over tax revenues, and over educational and social policy) and some of the anti-state arguments lose their force. Does populism wither away, to be replaced by a spectrum of parties, or is it consolidated along with a regional identity and mobilisation around xenophobia?

The related economic question is the extent to which we can understand the economic processes and diversity of contemporary Europe using a purely territorial model. Certainly there are rich and poor regions, and some of the differences are spectacular, but there are also striking differences between high-technology centres and abandoned mountain villages in southern France or northern Italy. As Narotzky remarks, the centralisation of ownership and power can be combined with the decentralisation of production, so that the old territorial models of centre and periphery – which fired up, for example, Occitanian activists – become less accurate. If local and regional identities become the dominant frame for political mobilisation, how do you develop a politics which will address those centres of power and control? Those on the right who think that the plight of small shopkeepers and businesses will be resolved by unfettered neo-liberalism will presumably get what they deserve. Those on the left who see a new terrain for mobilisation, combining the local and the global, have a much harder task.

9   CONCLUSION

This book has documented the mobilisation of groups to achieve a radical re-ordering of society by methods involving direct, often violent, confrontation with the existing state. Generally they were also radical in their effect on social relations, creating polarised identities which denied social gradations or the existence of overlapping and cross-cutting forms of cultural practice. There is also a series of contrasts, at least at a first level of generalisation. In relation to social space, class movements enacted a struggle which was pervasive throughout a given society; nationalist movements struggled for territorial separation and the defeat of the enemy within. In relation to time, class movements were future-oriented in their attempt to establish a new kind of society, and ways of being which had never existed before. Nationalist movements, built around identities established through historical continuity, offered the reconstitution of ways of being which had been lost. In this conclusion I shall amplify and qualify these generalisations, looking in turn at the issue of identity, political strategies and the social transformations within which these movements emerged. It will be a selective and open-ended conclusion, concentrating on a few issues which might stimulate reflections and comparisons from others.

POLITICS AND IDENTITY

I have stressed that class movements articulate identities, but this does not mean that I wish to assimilate class to the category of 'identity politics', as currently construed. This would be an unhelpful move on a number of accounts. Class, whether understood as a socio-economic category or as a political subject, does not fit the model of 'rights-based' politics, of groups struggling for recognition, or celebrating their difference. Coole (1996: 24) has argued that the 'hegemonic status of the discourses of difference among those who lay claim to political radicalism' has led to a silence about class. In fact, as Grossberg suggests (1996: 90), even the upsurge of studies dealing with multiple identities, which claim to give parity of treatment to 'race, class and gender', very rarely fit class into the frame. It tends to disappear after a token appearance on the first page (see also Ortner 1998: 1). A second issue is the way in which the study of 'identity politics', within cultural studies, has been dominated by a concern with the construction of identity, based on the study of texts, and drawing theoretical insight from work analysing psychological rather than sociological processes. The theoretical work is sophisticated and challenging, exploring, for example, the way in which identity is established through the experience of loss and rupture – a theme which starts with Freud and is picked up by writers from Hall to Passerini. However, models derived from individual experience provide only a limited blueprint for the analysis of political movements, and although intellectuals have stretched notions of 'reading' and 'writing' to cover more and more aspects of life, there are limits to the model of 'texts' in trying to understand the dynamics of social exclusion, mobilisation and action.

There are conflicting strands within this field of study. Much theoretical work has gone into deconstructing essentialist notions of identity based on sameness, replacing them with a conception of identity as multiple, plural, or hybrid, based on difference, and emerging through a complex process which may include internalisation of the 'other' and externalisation of the self. The shift from essentialist to non-essentialist conceptions of identity is sometimes presented as a long evolution in theory, working its way through in psychology, philosophy, linguistics. At other points the shift is presented as a 'real-world' global and epochal move from the homogenising collective action and institutional politics characteristic of modernity towards post-industrial, post-colonial or post-modern social forms. If we then ask whether it is true that people have more identities, or more complex identities, than they did a hundred years ago, we raise the suspicion that the term identity itself is not stable in this analysis. It moves, in often unacknowledged ways, between a general usage emerging out of social organisation and practice (such as those of gender, locality, occupation or religion) and that of the 'master narratives' of class or nation.

The most open and stimulating of these accounts recognise that it may not be very convincing to tell the history of the twentieth century in terms of a move from all-encompassing totalities to fragmentation and complexity. Raymond Williams (1985) suggested that every generation believed the organic, integrated community was found more completely in the past. Hall (sometimes) follows him in warning us to be careful about assumptions that once upon a time the great collective social identities, like the working class, were homogenous and fully formed

(Hall 1991: 46). The evidence from places like Sesto San Giovanni indicates that working-class movements were heterogeneous and aware of it, making rather than made, their unity depending on a forward momentum. At the same time the contemporary world is still full of politics based on essentialism – whether the institutional practices of nation-states, those found in many rights-based identity politics, or in the more conservative social movements which current theory tends to marginalise. It is also important to note that the politics which expresses non-essentialised conceptions of identity, either of multiple selves or hybridity, is of a different kind from the movements explored in this book. Hall (1991: 57) suggests that this second kind of identity politics develops primarily in the cultural field, challenging and discomforting dominant representations through the exploration of multiple difference. He also suggests that the move from an essentialised to a non-essentialised politics of identity is associated with a shift from what Gramsci called the 'politics of movement' – of mass, co-ordinated activity – to 'the politics of position', of more protracted and dispersed struggles. The shift is portrayed as historical and irreversible (Hall 1996: 427), deriving from fundamental political and cultural transformations, including the growth of civil society.

These arguments about identity politics take us straight into historical and geographical generalisations about the differences between traditional and modern societies, or between eastern and western Europe. I want to put these on one side for the moment, and suggest that essentialist and non-essentialist discourses co-exist within contemporary cultures. I will concentrate instead on the issue of identity in the more restricted field of political movements. It seems to me that a strong element of essentialism, however theoretically unfashionable, is a necessary part of these movements. Grossberg (1996: 89) summarises essentialised identities as those 'defined by either a common origin, or a common structure of experience or both. Basically the struggle over representations of identity here takes the form of offering one fully constituted, separate and distinct identity in place of another.' This kind of politics does not obliterate other kinds of identity, but it does subordinate them, generating a view that a person's most important or essential experience is derived from membership of a class or a nation, and conversely that these provide the core framework for interpreting history. This identity is articulated through a narrative which establishes who 'we' are through two axes, one of which is biographical and diachronic, while the other is oppositional and synchronic. These work together, though in different movements and in different contexts one may predominate over the other. This schema is an alternative to the rigid essentialist/non-essentialist distinction, and seems more appropriate for the study of political movements. I will say a little more about it at a formal level before returning to substantive issues.

OF BATTLES AND BEGETTING

Narratives are stories told. They range from the more formal, authoritative written histories of a people, of a movement, or of an epic leader, through to the emblematic event, evoked in more ephemeral writing or at gatherings, fragments of 'folk-history' (*historiae*), or lessons which encapsulate a defining moment. The narratives unfold in time, and the relationship between past, present and future creates a particular 'semantics of time'. Therborn (1995: 4) has written, 'Modernity ... can be defined as an epoch turned to the future, conceived as likely to be different from and possibly better than the present and the past. The contrast between the past and the future directs modernity's "semantics of time".' It is precisely in this sense that Le Pen and the other leaders discussed by Holmes (see Chapter 8) are 'anti-modernity' in their political visions, distancing them from this conception of progress. The point can be extended. Some movements, especially nationalist ones, have narratives which stress a foundation moment, and move through a succession of events which establish identity through continuity: a people remains 'true to itself' if it stays close to its origins and foundation; authenticity is associated with the past. Others, especially class movements, are oriented towards a future which is either a continuation of current processes or a transformative revolution – possibly a messianic end of history, so that the present is incorporated into that future, and a movement remains 'true to itself' if it stays on a road going forward.

These aspects of an identity discourse are directed 'inwards'; they are about unity and solidarity; when working smoothly, the historical narrative is redolent of the chapters of 'begetting' found in Genesis. They are also combined with the symbolic and ritual elements celebrated when people are 'amongst their own' and which we associate with unity in many contexts. But this is also a contested terrain, where people compete for leadership and make accusations that a movement has been betrayed – accusations which are again generated out of the 'semantics of time' characteristic of each movement: either that it has lost its roots and its links to the past, or lost its direction and its links to the future. These, then, are part of the 'internal politics' of a movement, fought out in terms of who best represents and can speak for authentic Basque culture or the interests of the working class in a movement prone to schism. This is how the book started, with the struggle over possession of the red flag of Lassalle.

But, as already indicated, this is only half the story: identity is also established through difference. This perception, at an abstract or conceptual level, was developed in theoretical work on the arbitrary nature of the linguistic sign, but in the study of identity the notion of 'difference' has to be widened to include a political as well as a conceptual level. This feature is most important in class movements, since their whole purpose is not to defend some free-standing essence, but to mobilise in opposition to a system of domination and to create a classless society. Class is conceived as a relationship, and the discourse includes an account of how society itself came to be divided. At a more practical level we have seen how broad-based class movements, like the Italian Communist Party, have appealed to the interests of all those who shared an opposition to monopoly capital.

In the case of national identity, we will often find that their histories contain a mixture of elements stressing opposition and continuity, battles and begetting. This dual character is revealed in the parts of national histories which deal with foundation – the moment in time when they became who they are. Foundation often involves a 'myth of disjunction' (Denich 1993) such as the religious conversion of an existing people (as with the Irish), or the migration of a people to occupy their present land (as in much of eastern Europe), though the Basques claimed to be autochtonous, and merged their history with biblical myths. But there is usually a second foundation moment, which focuses attention on the relations of dominance or subordination which mark a nation's history: the loss of autonomy to Castilian Spain, the defeat of the Serbs by the Turks in Kosovo. This is the 'external politics' of a fight against the Spanish state to achieve autonomy, or to reposition the Serbian people within a Yugoslav federation. In the case of some contemporary racist movements there may, for various strategic reasons, be considerable silence about what 'we' share, and self-definition is generated to a large extent out of opposition: who we are against.

Other aspects of this process can only be understood by moving away from identity-as-narrative, and putting these mobilising discourses back into a social context. We have seen that these movements operate at different levels, widening upwards from the face-to-face local interaction in a Bosnian village or an Italian factory. Articulating the connections means building 'imagined communities' out of real communities, or at least out of the immediacy of everyday experience. This has often involved the use of homology – that is, through stressing that what people share is a common structural position, of difference or subordination. I first used the term homology to discuss an aspect of Basque nationalism. There is of course a century-old political tradition of defining the Basque people in terms of essences – shared blood, language or culture. But an increasingly prominent second strand was to define the nation as all those within the territory who found themselves in opposition to the Spanish. Here and elsewhere, an authoritative discourse can establish identity between people who never meet by creating a powerful interpretation of the relations of difference and opposition which structure their everyday lives.

So identity is not only a narrative, it is part of social practice. In Chapter 5, summarising class movements, we saw the different ways people were brought together in grass-roots organisations, and the way the lines of class division were polarised by political actions such as the general strike. The two chapters dealing with nationalism and identity drew on work which moved away from mapping 'bits of culture' onto bounded groups, and looked at the dynamics of ethnicity in relation to social practice. The people of Bilbao, Sarajevo and their rural hinterlands are engaged in many relations, including those established through kinship, neighbourhood, work, education and religion. Sociologically these relationships may overlap, to create more totalising and bounded worlds, or they may cut across each other, so that the natal village, the residential neighbourhood and the workplace are worlds apart. The social worlds may merge or separate, and these patterns are restructured over time as the larger processes work their way through: migration, social mobility, growth and recession, and the effects of various state policies.

These sociological processes obviously affect identity and mobilisation, but I do not want to suggest that this happens in a linear or mechanical way. In times of rapid change and dislocation one set of relations may represent security and continuity, the home or homeland, which makes people what they are, through and through, in whatever context they find themselves. Once again, though, in order to understand how aspects of culture and performance become integrated into an ethnic identity, we need to look at the interplay between cultural differences and social boundaries. This includes the process I have referred to by the ungainly phrase 'the ethnicisation of social relations', and in the case studies it began with the analysis of particular social forms – the neighbourhood groups and voluntary associations of Basque society, the *nacije* of Bosnia. These groupings, generated through processes of inclusion and exclusion, and with their own moral vocabulary of trust, reciprocity and betrayal, were essential to the building of a wider political movement. Both regions have been socially polarised through violence. I have followed other commentators in suggesting that violence should not be seen as an innate consequence of ethnic difference, but as a political strategy which consolidates internal leadership and hardens ethnic boundaries by obliterating social interaction across them.

MODERNITY, OR WHAT TIME IS IT?

One of the recurring themes in the political movements we have examined is the fate of rural society, whether that involved labourers, share-croppers or small farmers. We have also found visions of autonomous and egalitarian rural communities embedded in political movements as diverse as Basque nationalism, Andalusian anarchism and east European populism. The prominence of rural themes derives partly from the priorities of the first generation of anthropologists, but it is also a reflection of the numerical importance of this population and the political importance of their mobilisation and dispersal. Hobsbawm (1994) has suggested that the most important long-term feature of the twentieth century is that, for the first time in human history, urban dwellers began to outnumber those who gained their livelihood from producing food. It is easy to forget how recent this is: even in Europe, one of the most industrialised corners of the globe, it only happened towards the middle of the century. Table 9.1 gives some census figures which show that in 1950 agriculture employed a quarter of the population in France and Germany, between 40 and 50 per cent in southern Europe, and between 50 and 75 per cent in eastern Europe. If we take into account the migration process and the generational lag, then even for some time after 1950 the majority of adults in Europe had direct experience of a rural world.

One recent contribution to the study of nationalism has very explicitly drawn attention to the importance of this rural world and its recent collapse. Tom Nairn's argument in *The Curse of Rurality* (1997) for the most part converges with that of Gellner, in the connections between modernisation, industrial society and nationalism. However, he suggests that the majority of ethno-nationalist conflicts seem to occur in pre-dominantly rural situations (Nairn 1997: 90) and, as a result, he gives a significantly different emphasis to the historical transition. While most accounts had stressed the emergence of a new social order – the cultural forms and rationality of industrial society – Nairn places centre stage the experience of those who went through the transition, and particularly of two groups. One is made up of a certain kind of urban intellectuals (such as those of the Basque provinces), 'who seek to mobilise lost-world psychology in order to build a new world, that of the modern nation-state' (Nairn 1997: 91). The other is the rural migrants themselves, who look backwards as much as forwards. In the disruption and dislocation of the industrial transition, both are haunted by the presence or the memory of a rooted, stable, rural world.

Table 9.1   The demise of rural Europe

Percentage of working population in agriculture 1910–80

	1910	1930	1950	1960	1980
North-Western Europe					
UK	9	6	5	4	3
France	41	36	27	22	8
Germany	37	29			
FRG			23	14	4
GDR			27	18	10
Southern Europe					
Italy	55	47	42	31	11
Spain	56		50	42	14
Portugal	57		49	44	28
Greece	50	54	51	56	37
Eastern Europe					
Poland	77	66	54	48	31
Czechoslovakia	40	37	39	26	11
Hungary	58	53	51	37	20
Yugoslavia	82	78	71	63	29
Romania	80	77	74	67	29
Bulgaria	82	80	65	56	37

Source: G. Ambrosius and W. Hubbard, *A Social and Economic History of Twentieth-Century Europe*, Cambridge, Mass., 1989: 58–9.

This cultural strand is so strong that Nairn suggests 'ethnic nationalism is in essence a peasantry transmuted, at least in ideal terms, into a nation'. Even in France, the home of enlightened civic nationalism, there was always another powerful strand in the construction of French nationality, one epitomised in the figure of Nicolas Chauvin, the ploughman-soldier (Nairn 1997: 103), who would emerge from his rural darkness to fight for his country and articulate the connections between blood, soil and xenophobia. Nairn argues that this rural world is a 'curse', in the sense that its haunting presence is *one* of the factors which gives virulence to ethno-nationalist conflict. He develops this analysis primarily with reference to Cambodia; in fact the article is built around an extensive review of Kiernan's book (1996) on the Pol Pot regime, arguing that Cold War dogmatics had obscured the fact in its murderous attempt to construct an ethnically pure rural society: 'The Cambodian

Hell was more truly an aberration of nationalist development than of socialism' (Nairn 1997: 92). Nairn suggests other parallel processes in Rwanda and Yugoslavia, and in less dramatic but equally murderous contexts like Northern Ireland and the Basque country.

Nairn's perspective is summarised in a striking phrase, 'Modernisation involves passage through something like a colossal mill-race, in which a multigenerational struggle between the rural past and the urban–industrial future is fought out ... where one global mode of existence perishes to make way for a successor' (Nairn 1997: 104). When did all this happen? It is still going on; the curse, the turmoil, are still with us. Even in industrial heartlands the rural past is much more recent than people realise. 'The kind of remaking which features in modern nationalism is not creation *ex nihilo*, but a reformulation constrained by determinate parameters of that past' (Nairn 1997: 104). The 'intense emotionality and violence' (Nairn 1997: 106) of nationalism is derived from those rural roots. Gellner gave too much emphasis to modernity and the new; his argument foreclosed too abruptly on the past. We may be living in a much earlier period of the modernisation process than intellectuals realise.

The thunder of the long collapse [of an ancient rural world] is still by far the loudest sound in all our ears. From Frank McLuhan to Baudrillard, theorists have sought to discern electronic post-modernity through the clouds of dust ... Many intellectuals go on believing it is much later than most people and politicians think. (Nairn 1997: 122)

In this article, and in earlier work (Nairn 1981) which introduced the symbol of Janus into our understanding of nationalism, Nairn looks at the reaction of populations living through rapid change – change which is, or is perceived to be, destructive of social forms which are rooted, particularistic, organic. We first met a version of this, paradoxically, in the context of class movements, in the loss and alienation of artisans recruited into a factory system, and in the collective memory of anarchists, and the way in which they could be harnessed to a vision of the future. However, the most striking narratives of loss – the backward-looking face of the Janus – were those we encountered in the nationalist movement. The nation had rural roots which were celebrated and revisited, in German, or Basque or Serb public cultures, and this provided the particularistic 'bite' (Nairn 1997: 104) that made people different. Rural 'ethnic' particularity could be incorporated into a 'forward-looking' vision, and into a state-building movement which embraced economic development, citizenship and the rule of law. This is how nationalism yokes cultural particularity to universalising modernity. However, for

Nairn the danger is that rural ethnicity remains forever backward-looking. Unlike the rural Arcadia beloved of Serb or Basque nationalists, his own representation of the rural world is an *oubliette* from which emerges all the dark forces of contemporary nationalism. Peasants are to Frenchmen as tradition is to modernity, and as emotion is to reason.

None of these images is very helpful in understanding the politics of nationalist mobilisation. Peasants should not be invoked to explain the passions of nationalism – the uncomfortable survival of emotion in the age of reason. Peasant societies are neither heaven nor hell, and they are certainly not timeless. They do in many cases have a strong tendency towards social closure, and once those boundaries become synonymous with those of the nation, as we saw in the Basque provinces and in Bosnia, they become very powerful and resistant. But do peasants loom out of the past to haunt the nation? The way in which peasant culture becomes national culture – the way rurality is invoked in nationalist discourse and peasants themselves are incorporated into nationalist movements – all originate with social forces outside the peasant world.

This brings us to the more difficult conceptual question about the 'semantics of time' and how we represent social change. Nairn's image of the mill-race, and the multi-generational struggle between the rural past and the urban–industrial future, is a striking image of tumult, and a reminder that there is an interaction between rural and urban. However, a mill-race only takes people in one direction. Gribaudi's (1987) study of the Torinese working class, or Holmes' (1989) study of the 'worker-peasants' of Friuli document a two-way movement, and show that people, over their working lives, moved back and forth between factory employment and their agricultural holdings. The evidence from Yugoslavia, when an industrialising economy imploded, shows in a more dramatic way that modernisation (and 'modernity') are reversible. In Yugoslavia and elsewhere, people were recycled from rural to urban and back again, sucked in and spat out. In each case the rural world comes to represent stability and continuity – the roots to which one returns in periods of crisis.

The issue of representation needs one last comment. The idea that peasant worlds are static is an urban myth. Many of the crucial features of peasant society only emerged after the abolition of feudal relations in the nineteenth century, and even after they were penetrated by market relations (or by socialist reorganisation) in the twentieth century. They may have changed at a different rhythm to the city (and the city was largely indifferent to what was happening), but in all the countries of southern and eastern Europe they also went through periods of extraordinary transformation, economically and culturally. Tradition, as Hall

(1992) and others have remarked, is the word given to what we are losing. However, even if, for the sake of argument, we describe the rural world as static, experiences are cumulative, and an urban worker who has been temporarily or permanently 'recycled' cannot simply return home, or return to the past. Such individuals may reveal themselves as fervently attached to a particular conception of work, property and household continuity, but it will have a dimension which is missing for somebody who has never left, or lost their livelihood. In Bourdieu's terms, all the things that go without saying have now been said. The heightened moral discourse around households and homelands which is such a striking feature of many nationalist movements is not in any simple sense a step backwards; nor does it involve atavistic figures who come back to haunt us. It is more likely to derive from the co-existence of the knowledge of different worlds, to be 'after-modernity' rather than a return to the past.

POLITICS AND SOCIAL PROCESS

These political movements were both a reaction to the large-scale trans-formations of society in the last century, and themselves part of that history, shaping the direction in which society moved. Much energy has gone into explaining why political movements occurred in the place that they did: factor x, (or x, y and z acting together) produced class struggle or nationalism. But European history is resistant to these kinds of gen-eralisation: central Italy goes through long periods of class mobilisation in the absence of a proletariat; a powerful explanation of Catalan nationalism is built around factors which are missing in the Basque provinces. The intensity of ethnic politics does not correlate with the degree of cultural diversity: Freud once said that the Balkans were char-acterised by the narcissism of marginal differences (Sahlins 1999: 413), and we can accept an underlying truth in that even if we would con-ceptualise it differently. Wiegandt, in an intelligent review of eight case studies designed to reveal the reasons for the presence or absence of ethnic mobilisation amongst rural populations, ends up concluding, in effect, that it depended on historically specific factors affecting the *interaction* between states and their minorities (Wiegandt 1993: 318). The most powerful theoretical frameworks for analysing political history employ concepts which are so generalised – capitalism, industrialism, modernisation – that they themselves have to be unpacked before they can have much purchase on what is happening in any specific case. This is not to deny the value or stimulus that can come from the attempts to establish very broad connections, but the next section will be a selective look at some of the processes which have emerged in the case studies, and in unpacking the generalising concepts.

Capitalism is an extraordinarily dynamic system in the speed of its expansion, and in the range of its transformations. As a result we can easily write as though this way of organising production is a monolithic and unstoppable force of nature, and in doing so end up subscribing to another kind of teleology. We need to keep in sight the global reach of capitalism and the enormous potential for accumulation inherent in this way of organising an economy, but we also need to remember that this potential is never realised in the abstract. It is realised, and society is re-ordered, through a combination of economic, technological and political circumstances. It happens through collective agency, combinations of economic and political actors who need not just to realise a profit but to transform the environment in which business is conducted. Together they promote or obstruct land reforms, build railways or mobile phone networks, switch energy policies, deregulate finance or create single markets. The dazzling rise of manufacturing industry in early twentieth-century Milan required (in no particular order) local entrepreneurs, hydro-electric power, German bankers, American knowhow and an Italian government keen to develop a national armaments industry. The equally dazzling rise of Bilbao a generation earlier, or the industrial districts of the 'Third Italy' a generation later, were the product of a similarly varied combination of interests and strategies.

All this suggests the importance of historical and geographical variations, and the last chapter noted that capitalism does not arrive in one hit. There are periods of accelerating change, when governments widen private property rights, abolish internal tariffs, create the communications infrastructure for a national market, and when there is a major investment in new manufactures. Concentrating on these dramatic periods can leave the impression that a society is from then on uniformly 'capitalist', moving inexorably forward. In reality, as innumerable studies have shown, this is not a smooth or unilinear process, and it is precisely the uneven character of subsequent development which leads to some forms of political mobilisation.

It is uneven firstly because transformations do not happen uniformly across the economy, but move from sector to sector, sometimes very slowly. The establishment of capitalist forms of organisation in Italy can be dated to the industrial boom and liberal reforms of the Risorgimento (or even to the Renaissance cities), but 80 years after those reforms there were still substantial parts of agriculture only partially incorporated into the market and farmers who were still coming to terms with the huge cultural transformations intrinsic to a market rationality (see Pratt

1994). At the end of the twentieth century (but before e-commerce) a revolution in retailing cut swathes through the small family businesses in the commercial sector, the backbone of many urban centres. This is the dynamism of capitalist systems, creating new products and new technologies, reorganising production and distribution so that new 'profit zones' emerge, often to displace household-based enterprises (Gudeman and Rivera 1990). There is a constant interaction with domestic labour, so that much food preparation now takes place in factories, while households are persuaded to assemble their own furniture. Changing returns to capital in an increasingly deregulated world also mean that whole sectors of industry can disappear or relocate to other parts of the globe. Sesto San Giovanni, for decades one of the innovative centres of Italian engineering, is in the year 2000 an industrial graveyard seeking funds for reconversion.

The unevenness of capitalist development is also regional. Some places have a history of investment, and have developed dense infrastructures, a skilled labour force and an administrative and education system which give them a competitive advantage. This is the case in Silicon Valley or the industrial districts of Italy (Bagnasco 1977; Sabel 1982), which are seen as sucking in capital investment and labour from poorer regions, accentuating regional disparities in an economic version of osmosis. There are disagreements about the factors which give such districts a competitive advantage, and we can certainly overstate the stability of any given regional pattern. Few looking at Andalusia's 'hungry coast', one of the poorest places in Europe in the 1940s, would have imagined the Costa del Sol 50 years later. Nevertheless it is clear that in economic terms territory is not homogeneous space, and that massive regional disparities exist: using European Union classifications some regions have *average* per capita incomes six times higher than others (Dunford 1996), and the disparity increases when we look to those countries seeking accession. Given that the EU already incorporates almost all the most prosperous parts of the continent, the economic gradient between those inside and those outside steepens with each enlargement.

A final kind of unevenness derives from the changing organisation of production in capitalist economies. With evidence accumulated in the century since Marx's death, few now believe that capitalism advances only through dispossession and a swelling multitude of wage-labourers. Agriculture was always out of step with this prognosis: the employment of wage-labourers boomed in southern Europe at the end of the nineteenth century, and for a long time these labourers outnumbered those in industry. But this was based on traditional labour-intensive technology, and the mechanisation of agriculture led to the dominance of family farming. Textile production was once mostly in the hands of artisans, but was then re-organised in factories driven by steam and electrical power, and now sections of it are fragmented back into household forms. An evolving combination of factors, from technological innovation and fiscal policy to changes in the pattern of consumer demand, mean that production processes are constantly being reorganised and involve a spectrum of economic actors. Freelance designers and consultants, wage-labourers, co-operatives of small producers, domestic out-workers may all be involved in one finished product, while in the wider society we have to add in another large stratum: the poorest, often called the underclass, who derive no income from employment at all.

We may identify a long-term dynamic within the economy, 'the increased accumulation of capital through the appropriation of surplus value' (Narotzky 1997: 217), but this does not happen through a linear process, and appropriation can be realised in a variety of ways. The result is considerable variation in the way production is organised, over time and between economic sectors. With this in mind, we can go back to the issue of class, which as a term refers to both economic strata and political actors. The case studies dealt with classes as collective political actors, as they emerged in parts of southern Europe marked by social polarisation, usually (but not invariably) the result of the formation of a critical mass of wage-labourers in large enterprises in agriculture or industry. These classes formed the core of movements which foresaw either the spread of wage-labour relations to the entire economy and the revolutionary seizure and transformation of the state, or the creation of a new society based on local autonomy and equality. There were many other dimensions to these movements – not least, in the first case, a strong belief in the unfolding of historical laws and the inevitability of socialism. Neither of these programmes now looks remotely plausible, and their proponents occupy a very marginal position in political life, but the 'death of class' does not mean that capitalism has lost its dynamism, a dynamism that comes precisely from the creation of disparities. This happens globally between north and south, and between the territorial regions of the industrialised states. In Italy the economic gap between north and south has increased during the last 50 years of state intervention. Disparities are also internal to a society; stratification can be analysed in a number of ways, and class remains the main term used to refer to these strata and to structured economic inequality (Coole 1996: 17). In that sense class did not die with class movements; economic disparities have continued within industrialised societies, and increased in those which embraced neo-liberalism (Krugman 1994). These disparities shape people's lives in many ways, but 'class politics' in this context refers to a very different process from the kind of polarisation we have examined. One statistically defined stratum does not mobilise against the others.

Nevertheless, political mobilisation around structured inequalities does continue. Naomi Klein is one of those who have documented some of the astonishing accumulation of capital by transnational corporations in the last 20 years. Here too we should not see this recent expansion as natural, but as an historical phenomenon generated out of a combination of processes and strategies, not least deregulation. Klein concentrates on the Dutch auction over labour costs and workplace discipline which has led to the transfer of manufacturing from the old industrial centres to the export zones of Mexico, the Philippines and southern China. Her comment, 'In this new globalized context, the victories of identity politics have amounted to a rearranging of the furniture while the house burnt down' (Klein 2000: 123), is an attack on the loss of radical economic analyses in movements built on identity politics, but it cannot be taken as an argument for the return of earlier forms of class mobilisation.

Her work documents some of the new configurations of concentration and dispersal. The vast financial power of many corporations makes controlling them through 'normal' political means – political parties and state legislation – virtually impossible, since they can force open markets, dictate to governments, and if necessary buy them off. The fragmentation and dispersal of the production system make it harder to achieve workplace solidarity, and create new patterns of dependency. A second strand in her study is privatisation – of services, utilities, property, public spaces and scientific knowledge. Klein then explores the strengths and weaknesses of new political movements, including environmental and human rights activists, union organisers in north and south, fair traders, consumer groups and culture jammers. Many of these are mobilised around a strong sense of locality, and some acknowledge an older anarchist agenda in their strategies. We saw in the last chapter that the politics of the local, and of territorial identities, are of growing importance, but also that they have many faces. When not linked to a concern with global processes and some measure of co-ordination, they can become largely conservative and exclusionary. Conversely, from Klein's perspective, there are limits to 'internet' activism and international rallies if they are not rooted in local campaigns. So there is mobilisation, especially around the twin themes of corporate power and privatisation, but it is not class mobilisation, and it is certainly not a Leninist mobilisation, much to the disappointment of media commentators looking for leaders and ideologies to criticise. Most of the other issues are still open: the substantive links between the various political agendas, the durability of the coalition, and the effectiveness of emerging political strategies. These will be key questions in twenty-first-century politics.

However, they will not be the only questions; others will arise from the changing political sovereignty of regions, nation-states and the European Union. A strong theme running through the case material has been the consolidation of states from the nineteenth century onwards as homogeneous spaces marked by territorial boundaries. The concept of *homogeneity* emerges in many discussions of the modern state (see Verdery 1994: 45) and can be seen as a way of taking forward Weber's analysis of rationality and the drive to consistency and efficiency (Gellner 1983: 20) in the ordering of many realms of modern society. Without going into debates about the priority of one social realm over another, we can mention some of the processes.

One dimension of this homogeneity is economic. Nationalists generally advocated protectionism against external competition, moving controls over the movement of goods to their frontiers, while removing all internal tariff barriers. This creation of a national market was further stimulated by massive investment in internal means of communication: roads, railways, telegraph. The leitmotif of policy was 'getting things moving', ripping out any internal obstructions, so that, in a favourite metaphor, the health and growth of the nation depended on movement, just like blood coursing through the body. This is the dynamic of a market economy: that the same forces operate throughout a territory even if, as we have already seen, the result is usually an increased disparity in the distribution of wealth. The rationality of socialist economies manifested another form of order in the planning and allocation systems which applied throughout state territory, drafting the people-as-workers into strategically decided production roles.

State practice assumes and generates uniform procedures in many other institutional fields: in the creation of a universal rule of law and the rights associated with citizenship, in the practice of public (and private) bureaucracies. It is not that such a system of government necessarily eliminates patronage or creates equity, it is that it eliminates the right to differential treatment on the part of particular sections of society, or localities which had existed in earlier regimes. In the context of government Verdery notes that

a homogenising policy creates the 'nation' as consisting of all those the state should administer, because they all ostensibly 'have something in common'. State subjects are most frequently encouraged to have 'in common' (besides their government) shared culture and/or 'ethnic' origin. (Verdery 1994: 45)

It is in this context that we encounter the most famous arguments of nationalism – the way the state creates linguistic uniformity through the standardisation of a vernacular language, and its use throughout state institutions, not least the newly compulsory state education system. Schooling, literacy and the development of the media in turn lead to the development of a national language, literature and history – a degree of cultural standardisation which is contested, partly mythical but also partly realised.

The point of this brief sketch is simply to draw attention to the complexity of these social transformations, and to the various dimensions that unfold at different speeds and interact in different ways in local contexts. The experiences described in this book were from those sections of society that were losing out in these transformations and sought a political solution, though we should not forget the millions who left southern Europe and migrated to the Americas. But there were many kinds of dislocation, and the result was a cross-cutting set of divisions, and often a variety of potential political programmes: not just an impoverished peasantry, but a Catholic, Euskardi-speaking peasantry; not just a wage-labourer but an agricultural labourer immersed in the anti-Catholic moral codes of the *pueblo*. Small farmers, craft workers, traders and other parts of the 'petty-bourgeoisie' did not simply lose their incomes, they lost their livelihoods and their property, and their skills became valueless. Their associations and networks were disbanded along with their political position within local society, as the new circuits of capital came to dominate. Older forms of knowledge and skills were devalued, and their previous distribution within society became irrelevant: this was a profound cultural revolution and became dramatic with the generalised introduction of schooling, so that the key processes of cultural acquisition were organised by the state and fell outside local control. There is a marked impact on social divisions, since literacy is a prerequisite for operating successfully in a widening market, and to move into the new professions; only the more slow-moving, local and 'informal' sectors of the economy can continue to operate without it. The consequences of this are even more dramatic if the language of the state is significantly different from the local vernacular, and this is one of best-researched aspects of European nationalism.

Hroch (1985: 177–91) and other scholars have devoted particular attention to the political options and strategies of the 'middle strata' comprising both the economic actors (artisans, traders, small-business owners) and the professionals or intellectuals (clergy, lawyers, teachers). Did they think that they would have a stake in the new kind of society emerging or did they oppose it? Were they united in their response, and with whom would they ally in opposition? In the Basque provinces, for example, where the emerging class divide and linguistic divides fell in different places, the situation was very fluid. In the end the majority of these urban middle strata of intellectuals and small businessmen embraced Basque nationalism and allied with the peasantry; in slightly different circumstances they could have been included in a Christian Democrat movement which championed household enterprises on a pan-Spanish basis. In the same period, but in a region with a different economic history, the Milanese artisans became a founding force in the Socialist Party. The rural populations of eastern Europe in the first half of the twentieth century were also active in, or co-opted by, a variety of political movements: populist–nationalism, green agrarian internationalism, and the red Communist International.

These comments suggest that the political configuration which emerges in the complexity of the nation-building process is highly variable. It depends on many factors, including the forces which dominate within the state and the linkages between centres and regions. In the context of nationalism, Heiberg (1989: 231–43) has pointed to the distinction between societies where local elites or notables were co-opted into the state machinery, and those where this did not happen. It also depends on timing: the gap between the destruction of existing modes of livelihood and the eventual arrival of new resources. For the other side to this 'open' economic and political space is the drive to create national economies through investment and protectionism for strategic economic sectors, since only the strong (like Britain in the nineteenth century) favoured free trade. These 'Fordist' economies of mass production and consumption were national in the sense that they were owned by and produced for nationals; autonomy (achieved without trade) was often valued since it reduced dependence, while foreign ownership was considered a threat to national interests. Taxation systems were the base for public expenditure; the levels of both were politically contested, but led overall to the development of public education and 'cradle-to-grave' welfare systems which were very important in the formation of citizenship and nationhood. So the state always had a multiple role, providing much of the infrastructure and regulatory system necessary for capitalist operations, but also as a potential mechanism for redistributing wealth between regions and sectors, counteracting the inegalitarian dynamic of the market (Wicker 1997).

This model of culturally homogeneous, economically autonomous, politically sovereign states was never completely achieved, but it is now being significantly eroded, the directions reversed. Globalisation is another overworked term. One of Klein's informants in the Philippines reminded her that it had always been part of a global society (Klein 2000: 439) – at least since the sixteenth century, when the Spaniards moving west met the Chinese moving east. But though many phenomena are not as new as is sometimes claimed, there has been over the last two decades a novel series of interlocking processes, including the development of information technology, deregulation of capital movements, trade liberalisation and new patterns of migration. The European Union has consolidated itself as a supranational entity, and become one of the poles in the new patterns of production and trade. Again it is possible to overestimate the scale of these changes, but their cumulative impact has been felt in all spheres: an increase in certain forms of cultural heterogeneity, a reduction in sovereignty, in the boundedness of national economies, and in the state's capacity to redistribute wealth, undermined by the ability of the largest corporations and the most successful professionals to avoid taxation by operating transnationally (Reich 1991).

If globalisation produces weak states, it does not necessarily weaken nationalism. All these processes generate reactions. Political forces in existing states can mobilise opposition to loss of power to Brussels, or over the growth of transnational migration, by reasserting the rights of nation-states. Often such movements obscure the economic processes which lead to loss of sovereignty, and stress the threat to the national way of life, represented in very traditional terms. The growth of regional disparities and the incapacity or unwillingness of governments to alleviate them can generate or reactivate ethno-nationalist movements within existing states, in both prosperous and disadvantaged regions. The strengthening of the European Union opens up a conceptual as well as an economic space for rethinking the existing map of nations, as we saw in the previous chapter. The collapse of communism in eastern Europe has also generated strongly nationalistic movements – not through unleashing sleeping giants freeze-dried by Stalinism, but as a consequence of the centralised allocation systems of communism itself and the chaotic circumstances of many 'transitions'. Nationalism and autocratic governments have emerged both in those societies which embarked on shock-therapy liberalisation, and in those which witnessed a retrenchment of state power (in the economy, media and military). Yugoslavia was a tragic example of what could go wrong in a region where renewed conflict around the relationship between states and minorities had led to a process of political reconfiguration (see Brubaker 1996).

Events in eastern Europe are sometimes dramatic and destructive, but reconfiguration is occurring throughout Europe, as the economic, political and cultural processes pull in opposite directions. In an increasingly unbounded world, some political parties and movements respond to this by advocating the strengthening of nation-states, even as supranational forms of government are consolidating. Others are moving to scale down the units of government, to reinforce democracy at the level of regions or 'sub-nations', and counteract the centralised and largely unaccountable powers of the European Union. Such developments have many virtues, but they do not necessarily solve the 'problem' of cultural heterogeneity; in fact they may be counterproductive in the development of a politics of cultural pluralism, and in themselves simply displace the wider problems of redistribution which were previously fought out at the level of the nation-state. Political movements address different parts of the configuration, and there is a variety of views about what a new map might look like – if indeed a map is not an outdated metaphor. The only consensus is that Europe is at the beginning of a period of accelerating change, and that the outcomes are uncertain.

We should be open-minded in analysing those outcomes. This book started by expressing dissatisfaction with the tendency to lock the study of class and nationalist movements into incommensurable paradigms, so that political action was fuelled by totally different kinds of people and motives. In one world we found economic categories of people driven by material interests, in the other cultural subjects consumed by passions; analysis of the first would start with the labour process, the second with identity and representation. Twentieth-century European history does not always fit into such categorisations. We saw that class mobilisation in Italy triggered competing versions of Italian identity, while civil wars in Spain, Finland and Greece have been seen as both national and class struggles.

I have highlighted the work of historians and social scientists who have implictly or explicitly broken with the desire to construct neat boxes for categorising political movements, and have revealed the complex interactions between different kinds of social change, together with the spectrum of political responses to them. This is not an argument that anything goes or anything is possible – revolutionary socialism continues to be a very different kind of animal from Lombard separatism. It is simply a reminder that similar processes may generate very different political movements. Economic restructuring, for example, has a major impact on territories and regions, and this is reflected, perhaps increasingly, in movements concerned with the defence of livelihoods. It is also a reminder that comparable (not identical) processes may be going on in very different movements. In the Introduction I suggested people living through periods of rapid change and dislocation forge an identity and an interpretation of history which makes their own experiences central in a narrative of how society should be, and forge a political strategy to make that happen. Political anthropology has a particular role in building an understanding of these movements, because it unpacks the abstractions of capitalism or nation-building, and provides a context which can bring into one framework the analysis of issues as diverse as the labour process and identity narratives. The book will have served its purpose if it helps others think about the different political configurations emerging and ask further questions about them.

REFERENCES

Anderson, B. (1983) *Imagined Communities* (London: Verso).

Anderson, P. (1980) *Arguments Within English Marxism* (London: Verso).

Anzulovic, B. (1998) *Heavenly Serbia* (London: Hurst).

Appadurai, A. (1986) 'Theory in anthropology: center and periphery', *Comparative Studies of Society and History*, vol. 29, pp. 356–61.

Aya, R. (1975) *The Missed Revolution. Papers on European and Mediterranean Societies*, no. 3 (Amsterdam: University of Amsterdam).

— (1990) *Rethinking Revolutions and Collective Violence* (Amsterdam: Het Spinhuis).

Bagnasco, A. (1977) *Tre Italie* (Bologna: Il Mulino).

Bagnasco, A. and Oberti, M. (1998) 'Italy: "*le trompe-l'oeil*" of regions', in Le Galès, P. and Lequesne, C. (eds) *Regions in Europe* (London: Routledge) pp. 150–65.

Baier, L. (1991) 'Farewell to regionalism', *Telos*, vol. 90, pp. 82–8.

Bailey, F.G. (ed.) (1971) *Gifts and Poison* (Oxford: Basil Blackwell).

— (ed.) (1973) *Debate and Compromise* (Oxford: Basil Blackwell).

Banac, I. (1996) 'Bosnian Muslims: from religious community to socialist nationhood and post-communist statehood, 1918–1992', in Pinson, M. (ed.) *Muslims of Bosnia-Herzegovina* (Cambridge, MA: Harvard University Press) pp. 129–53.

Bauman, Z. (1982) *Memories of Class* (London: Routledge and Kegan Paul).

Baumann, G. (1996) *Contesting Culture* (Cambridge: Cambridge University Press).

Bax, M. (1995) *Medjugorje: Religion, Politics and Violence in Rural Bosnia* (Amsterdam: Vitgeveril).

— (2000) 'Warlords, priests and the politics of ethnic cleansing: a case study from Bosnia-Hercegovina', *Ethnic and Racial Studies*, vol. 23, no. 1, pp. 16–35.

Beissinger, M. (1998) 'Nationalisms that bark and nationalisms that bite: Ernest Gellner and the substantiation of nations', in Hall, J. A. (ed.) *The State of the Nation* (Cambridge: Cambridge University Press) pp. 169–90.

Bell, D. H. (1986) *Sesto San Giovanni. Workers, Culture and Politics in an Italian Town, 1880–1922* (New Brunswick: Rutgers University Press).

Bell, J. D. (1977) *Peasants in Power* (Princeton: Princeton University Press).

Ben-Ami, S. (1992) 'Basque nationalism between archaism and modernity', in Reinharz, J. and Mosse, G. L. (eds) *The Impact of Western Nationalisms* (London: Sage).

Bideleux, R. and Jeffries, I. (eds) (1998) *A History of Eastern Europe* (London: Routledge).

Billig, M. (1995) *Banal Nationalism* (London: Sage).

Blackburn, R. (1999) 'Kosovo: the war of NATO expansion', *New Left Review*, vol. 235, May/June, pp. 107–23.

Boggs, C. (1986) *Social Movements and Political Power: Emerging Forms of Radicalism in the West* (Philadelphia: Temple University Press).

Bojicic, V. (1996) 'The disintegration of Yugoslavia: causes and consequences of dynamic inefficiency in semi-command economies', in Dyker, D. and Vejvoda, I. (eds) *Yugoslavia and After* (London: Longman) pp. 28–47.

Bonifazi, E. (1979) *Lotte Contadine nella Val d'Orcia* (Siena: Nuovo Corriere Senese).

Bougarel, X. (1996) 'Bosnia and Hercegovina: state and communitarianism', in Dyker, D. and Vejvoda, I. (eds) *Yugoslavia and After* (London: Longman) pp. 87–115.

Bourdieu, P. (1984) *Distinction* (London: Routledge and Kegan Paul).

Bowman, G. (1994) 'Xenophobia, fantasy and the nation: the logic of ethnic violence in former Yugoslavia', in Goddard, V. A., Llobera, J. R. and Shore, C. (eds) *The Anthropology of Europe* (Oxford: Berg) pp. 143–72.

Brenan, G. (1960) *The Spanish Labyrinth* (Cambridge: Cambridge University Press, first published 1943).

Bringa, T. (1993) 'Nationality categories, national identification and identity formation in "multinational" Bosnia', *Anthropology of East Europe Review*, vol. 11, nos. 1–2.

— (1996) *Being Muslim the Bosnian Way* (Princeton: Princeton University Press).

Brubaker, R. (1996) *Nationalism Reframed* (Cambridge: Cambridge University Press).

Chomsky, N. (1969) *American Power and the New Mandarins* (London: Chatto) pp. 23–129.

Clark, M. (1977) *Antonio Gramsci and the Revolution that Failed* (New Haven: Yale University Press).

Clark, R. P. (1979) *The Basques: The Franco Years and Beyond* (Reno: University of Nevada Press).

— (1984) *The Basque Insurgents: ETA, 1952–1980* (Madison: University of Wisconsin Press).

Clemente, P. (1980) *Mezzadri, Letterati e Padroni* (Palermo: Sellerio Editore).

— (1987) 'Mezzadri in lotta: tra l'effervescenza della ribellione e i tempi lunghi della storia rurale', in Clemente, P. (ed.) *Annali dell'Istituo 'A Cervi', 9* (Rome: Il Mulino, Istituto Cervi) pp. 285–305.

Collier, G. A. (1987) *Socialists of Rural Andalusia* (Stanford: Stanford University Press).

Conversi, D. (1997) *The Basques, the Catalans and Spain* (London: Hurst).

Coole, D. (1996) 'Is class a difference that makes a difference?', *Radical Philosophy*, vol. 77, pp. 17–25.

Corbin, J. (1993) *The Anarchist Passion* (Aldershot: Avebury).

Culhoun, C. J. (1988) 'The radicalism of tradition and the question of class struggle', in Taylor, M. (ed.) *Rationality and Revolution* (Cambridge: Cambridge University Press) pp. 129–75.

Davidson, A. (1982) *The Theory and Practice of Italian Communism* (London: Merlin Press).

Davis, J. (1977) *People of the Mediterranean* (London: Routledge and Kegan Paul).

De Luna, G. (ed.) (1994) *Figli di un Benessere Minore. La Lega 1979–1993* (Florence: La Nuova Italia Editrice).

Denich, B. (1993) 'Unmaking multi-ethnicity in Yugoslavia: metamorphosis observed', *Anthropology of East Europe Review*, vol. 11, nos. 1–2.

— (1994a) 'Dismembering Yugoslavia : nationalist ideologies and the symbolic revival of genocide', *American Ethnologist*, vol. 21, no. 2, pp. 367–90.

— (1994b) *Ethnic Nationalism: The Tragic Death of Yugoslavia* (Minnesota: University of Minnesota Press).

— (2000) 'Unmaking multiethnicity in Yugoslavia: media and metamorphosis', in Halpern, J. M. and Kideckel, D. A. (eds) *Neighbors at War* (Pennsylvania: Pennsylvania State University Press) pp. 39–55.

Despalatovic, E. (1993) 'Reflections on Croatia, 1960–1992', *Anthropology of East Europe Review*, vol. 11, nos. 1–2.

Douglass, W. A. (1988) 'A critique of recent trends in the analysis of ethnonationalism', *Ethnic and Racial Studies*, vol. 11, no. 2, pp. 192–206.

Dunford, M. (1996) 'Disparities in employment, productivity and output in the EU', *Regional Studies*, vol. 30, no. 4, pp. 339–57.

204   *Class, Nation and Identity*

Dyker, D. (1972) 'The ethnic Muslims of Bosnia: some basic socio-economic data', *The Slavonic and East European Review*, April, pp. 238–56.

— (1996) 'The degeneration of the Yugoslav Communist Party as a managing elite', in Dyker, D. and Vejvoda, I. (eds) *Yugoslavia and After* (London: Longman) pp. 48–64.

Dyker, D. and Vejvoda, I. (eds) (1996) *Yugoslavia and After* (London: Longman).

Escobar, A. (1992) 'Culture, practice and politics. Anthropology and the study of social movements', *Critique of Anthropology*, vol. 12, no. 4, pp. 395–432.

Fardon, R. (1990) *Localizing Strategies* (Edinburgh: Scottish Academic Press).

Foweraker, J. (1989) *Making Democracy in Spain: Grass-roots Struggle in the South 1955–1975* (Cambridge: Cambridge University Press).

Fraser, R. (1973) *The Pueblo* (London: Allen Lane).

Gellner, E. (1983) *Nations and Nationalism* (Oxford: Blackwell).

— (1997) 'Reply to critics', *New Left Review*, vol. 221, pp. 81–118.

Gill, D. (1983) 'Tuscan share-cropping in United Italy: the myth of class-collaboration destroyed', *Journal of Peasant Studies*, vol. 4, 146–69.

Gilmore, D. (1980) *The People of the Plains: Class and Community in Andalusia* (New York: Columbia University Press).

Ginsborg, P. (1990) *A History of Contemporary Italy* (London: Penguin).

Giorgetti, G. (1974) *Contadini e Proprietari nell'Italia Moderna* (Turin: Einaudi).

Gledhill, J. (1994) *Power and its Disguises* (London: Pluto Press).

Glenny, M. (1996) *The Fall of Yugoslavia: The Third Balkan War* (London: Penguin).

Goddard, V. A., Llobera, J. R. and Shore, C. (eds) (1994) *The Anthropology of Europe* (Oxford: Berg).

Gowan, P. (1999) 'The NATO powers and the Balkan Tragedy', *New Left Review*, vol. 234, pp. 83–105.

Greenwood, D. (1976) *Unrewarding Wealth* (Cambridge: Cambridge University Press).

— (1977) 'Continuity in change : Spanish Basque ethnicity as a historical process', in Esman, M. J. (ed.) *Ethnic Conflict in the Western World* (Cornell: Cornell University Press) pp. 81–102.

Gribaudi, M. (1987) *Mondo Operaio e Mito Operaio* (Turin: Giulio Einaudi Editore).

Grillo, R. D. (1989) *Dominant Languages* (Cambridge: Cambridge University Press).

Grossberg, L. (1996) 'Identity and cultural studies: is that all there is?', in Hall, S. and du Gay, P. (eds) *Questions of Cultural Identity* (London: Sage) pp. 87–107.

Gudeman, S. and Rivera, A. (1990) *Conversations in Columbia* (Cambridge: Cambridge University Press).

Hadjimichalis, C. (1986) *Uneven Development and Regionalism* (London: Croom Helm).

Hall, J. A. (ed.) (1998) *The State of the Nation* (Cambridge: Cambridge University Press).

Hall, S. (1991) 'Old and new identities', in King, A. D. (ed.) *Culture, Globalization and the World-System* (Basingstoke: Macmillan) pp. 41–68.

— (1992) 'The question of cultural identity', in Hall, S., Held, D. and McGrew, T. (eds) *Modernity and its Futures* (Cambridge: Polity Press) pp. 273–326.

— (1996) 'Gramsci's relevance for the study of race and ethnicity', in Morley, D. and Chen, K. H. (eds) *Stuart Hall: Critical Dialogues in Cultural Studies* (London: Routledge) pp. 411–40.

Hall, S. and du Gay, P. (eds) (1996) *Questions of Cultural Identity* (London: Sage).

Hall, S., Lumley, R. and Mclennan, G. (1977) 'Politics and ideology: Gramsci', in Hall, S. (ed.) *On Ideology* (London: Hutchinson) pp. 45–76.

Halpern, J. M. (1993) 'Introduction: special issue: war among the Yugoslavs', *Anthropology of East Europe Review*, vol. 11, nos. 1–2.

Halpern, J. M. and Halpern, B. K. (1972) *A Serbian Village in Historical Perspective* (New York: Holt, Reinhart and Winston).

Halpern, J. M. and Kideckel, D. A. (eds) (2000) *Neighbors at War* (Pennsylvania: Pennsylvania State University Press).

Hann, C. (ed.) (2001) *Postsocialism* (London: Routledge).

Hayden, R. (1996) 'Imagined communities and real victims: self-determination and ethnic cleansing in Yugoslavia', *American Ethnologist*, vol. 23, no. x, pp. 783–801.

Heiberg, M. (1975) 'Insiders/outsiders: Basque nationalism', *Archive of European Sociology*, vol. 16, pp. 169–93.

— (1980) 'Basques, anti-Basques and the moral community', in Grillo, R. D. (ed.) *Nation and State in Western Europe* (London: Academic Press) pp. 45–60.

— (1989) *The Making of the Basque Nation* (Cambridge: Cambridge University Press).

Hobsbawm, E. J. (1959) *Primitive Rebels* (Manchester: Manchester University Press).

— (1984a) *Worlds of Labour*. (London: Weidenfeld and Nicolson).

— (1984b) 'Man and woman: images on the Left', in Hobsbawm, E. J. *Worlds of Labour* (London: Weidenfeld and Nicolson) pp. 66–102.

— (1994) *Age of Extremes* (London: Abacus).

Holmes, D. (1989) *Cultural Disenchantments* (Princeton: Princeton University Press).

— (2000) *Integral Europe: Fast-Capitalism, Multiculturalism, Neofascism* (Princeton: Princeton University Press).

Howell, D. (ed.) (1993) *Roots of Rural Ethnic Mobilisation* (Dartmouth: New York University Press).

Hroch, M. (1985) *Social Preconditions of National Revival in Europe* (Cambridge: Cambridge University Press).

— (1998) 'Real and constructed: the nature of the nation', in Hall, J. A. (ed.) *The State of the Nation* (Cambridge: Cambridge University Press) pp. 91–106.

Hudson, R. and Lewis, J. (eds) (1985) *Uneven Development in Southern Europe* (London: Methuen).

Jackson, G. D. (1966) *Comintern and Peasant in Eastern Europe 1919–1930* (New York: Columbia University Press).

Juaristi, J. (1998) *El Bucle Melancolico. Historias de Nacionalistas Vascos* (Madrid: Espasa).

Judah, T. (1997) *The Serbs* (New Haven: Yale University Press).

Just, R. (1989) 'Triumph of the ethnos', in Tonkin, E., McDonald, M. and Chapman, M. (eds) *History and Ethnicity ASA 27* (London: Routledge) pp. 71–88.

Kahn, J. S. (1989) 'Culture: demise or resurrection?', *Critique of Anthropology*, vol. 9, no. 2, pp. 5–25.

Kaldor, M. (1999) *New and Old Wars* (Oxford: Polity Press).

Kaplan, T. (1977) *Anarchists of Andalusia 1863–1903* (Princeton: Princeton University Press).

Kaye, H. J. and McClelland, K. (1990) *E.P. Thompson. Critical Perspectives* (Philadelphia: Temple University Press).

Kertzer, D. (1980) *Comrades and Christians* (Cambridge: Cambridge University Press); new edition (1990) (Prospect Heights, Ill: Waveland Press).

— (1996) *Politics and Symbols* (New Haven: Yale University Press).

Kielstra, N. (1985) 'The rural Languedoc: periphery to "relictual space"', in Hudson, R. and Lewis, J. (eds) *Uneven Development in Southern Europe* (London: Methuen) pp. 246–62.

Kiernan, B. (1996) *The Pol Pot Regime: Race, Power and Genocide in Cambodia under the Khmer Rouge, 1975–79* (New Haven: New Haven).

Kitching, G. (1985) 'Nationalism: the instrumental passion', *Capital and Class*, vol. 25, pp. 98–116.

Klein, N. (2000) *No Logo* (London: Flamingo).

Krugman, P. (1994) *Peddling Prosperity* (New York: W. W. Norton and Company).

Laitin, D. (1995) 'National revivals and violence', in *Archive of European Sociology*, vol. XXXVI, pp. 3–43.

Le Galès, P. and Lequesne, C. (eds) (1998) *Regions in Europe* (London: Routledge).

Lem, W. (1994) 'Class politics, cultural politics', *Critique of Anthropology*, vol. 14, no. 4, pp. 393–417.

—— (1999) *Cultivating Dissent: Work, Identity and Praxis in Rural Languedoc* (New York: State University of New York Press).

Levy, C. (1999) *Gramsci and the Anarchists* (Oxford: Berg).

Li Causi, L. (1993) *Il Partito a Noi ci ha Dato* (Siena: Laboratorio Etno-Antropologico).

Lockwood, W. G. (1975) *European Moslems: Economy and Ethnicity in Western Bosnia* (New York: Academic Press).

Lumley, R. (1990) *States of Emergency* (London: Verso).

Lyttleton, A. (1979) 'Landlords, peasants and the limits of liberalism', in Davis, J. A. (ed.) *Gramsci and Italy's Passive Revolution* (London: Croom Helm) pp. 104–35.

MacClancey, J. (1993) 'Biological Basques, sociologically speaking', in Chapman, M. (ed.) *Social and Biological Aspects of Ethnicity* (Oxford: Oxford University Press).

Malcolm, N. (1996) *Bosnia* (London: Macmillan).

—— (1998) *Kosovo* (London: Macmillan).

Malefakis, E. E. (1970) *Agrarian Reform and Peasant Revolution in Spain* (New Haven and London: Yale University Press).

Martinez-Alier, J. (1971) *Labourers and Landowners in Southern Spain* (London: Allen & Unwin).

May, E. (1997) '"Primitive rebels" in Spain: historians and the anarchist phenomenon', in Stradling, R., Newton, S. and Bates, D. (eds) *Conflict and Coexistence* (Cardiff: University of Wales Press) pp. 196–218.

Mazower, M. (1998) *Dark Continent* (London: Penguin).

Mintz, J. (1982) *The Anarchists of Casa Viejas* (Chicago: University of Chicago Press).

Mitrany, D. (1951) *Marx Against the Peasant* (Carolina: University of North Carolina Press).

Morley, D. and Chen, K. H. (eds) (1996) *Stuart Hall: Critical Dialogues in Cultural Studies* (London: Routledge).

Mursic, R. (2000) 'The Yugoslav dark side of humanity : a view from a Slovene blind spot', in Halpern, J. A. and Kideckel, D. A. (eds) *Neighbors at War* (Pennsylvania: Pennsylvania State University Press) pp. 56–77.

Nairn, T. (1981) *The Break-Up of Britain* (London: Verso) 2nd edition.

—— (1997) 'The curse of rurality: limits of modernisation theory', in Hall, J. A. (ed.) *The State of the Nation* (Cambridge: Cambridge University Press) pp. 107–34.

Narotzky, S. (1997) *New Directions in Economic Anthropology* (London: Pluto Press).

Olsen, M. K. G. (1993) 'Bridge on the Sava: ethnicity in Eastern Croatia 1981–1991', *Anthropology of East Europe Review*, vol. 11, nos. 1–2.

Ortner, S. (1998) 'Identities: the hidden life of class', *Journal of Anthropological Research*, vol. 54, no. 1, pp. 1–17.

Passerini, L. (1979) 'Work, ideology and consensus under Italian Fascism', *History Workshop Journal*, vol. 8, pp. 82–108.

—— (1987) *Fascism in Popular Memory* (Cambridge: Cambridge University Press).

—— (1996) *Autobiography of a Generation: Italy 1968* (Hanover: University Press of New England).

Payne, S. G. (1975) *Basque Nationalism* (Reno: University of Nevada Press).

Pazzagli, C. (1979) *Per la Storia Dell'Agricultura Toscana nei Secoli XIX e XX* (Turin: Einaudi).

Pazzagli, C., Cianferoni, R. and Anselmi, S. (eds) (1986) *I Mezzadri e La Democrazia in Italia* (Rome: Il Mulino).

Periccioli, A. I. (2001) *Giorni Belli e Difficili, L'Avventura di un Comunista* (Milan: Jaca Book).

Piccone, P. (1991) 'Federal Populism in Italy', *Telos*, vol. 90, pp. 3–18.

Pinson, M. (1996a) 'The Muslims of Bosnia-Herzegovina under Austro-Hungarian rule, 1878–1918', in Pinson, M. (ed.) *The Muslims of Bosnia-Herzegovina* (Cambridge, MA: Harvard University Press) pp. 84–128.

— (ed.) (1996b) *The Muslims of Bosnia Herzegovina* (Cambridge, MA: Harvard University Press) 2nd edition.

Pitt-Rivers, J. (1954) *The People of the Sierra* (Chicago: University of Chicago Press).

Portelli, A. (1990) 'Uchronic dreams', in Samuel, R. and Thompson, P. (eds) *The Myths We Live By* (London: Routledge) pp. 143–60.

Pratt, J.C. (1980) *The Mezzadria* (Amsterdam: Euromed Institute).

— (1986) *The Walled City* (Gottingen: Herodot).

— (1987) 'La ricerca antropologica e la mezzadria', in Clemente, P. (ed.) *Il Mondo a Meta. Sondaggi Antropologici sulla Mezzadria Classica* (Rome: Editrice Mulino) pp. 35–53.

— (1989) 'Some Italian Communists talking', in Grillo, R. D. (ed.) *Social Anthropology and the Politics of Language* (London: Routledge) pp. 176–92.

— (1994) *The Rationality of Rural Life* (Reading: Harwood Academic Press).

— (2001) 'Political identity and communication', in Cheles, L. and Sponza, L. (eds) *The Art of Persuasion* (Manchester: Manchester University Press) pp. 87–98.

Przeworski, A. (1977) 'Proletariat into a class: the process of class formation from Karl Kautsky's *The Class Formation* to recent controversies', *Politics and Society*, vol. 7, no. 4, pp. 343–401.

Radosevic, S. (1996) 'The collapse of Yugoslavia: between chance and necessity', in Dyker, D. and Vejvoda, I. (eds) *Yugoslavia and After* (London: Longman) pp. 65–86.

Ramet, S. P. (1996) 'Nationalism and the "idiocy" of the countryside: the case of Serbia', *Ethnic and Racial Studies*, vol. 19, no. 1, pp. 70–87.

Reich, R.B. (1991) *The Work of Nations* (London: Simon and Schuster).

Ruzza, C. and Schmidtke, O. (1991) 'The making of the Lombard League', *Telos*, vol. 90, Spring, pp. 57–70.

— (1993) 'Roots of success of the *Lega Lombarda*: mobilisation dynamics and the media', *West European Politics*, vol. 16, pp. 1–23.

Sabel, C. (1982) *Work and Politics* (Cambridge: Cambridge University Press).

Sahlins, M. (1999) 'Two or three things I know about culture', *Journal of the Royal Anthropological Institute*, vol. 5, no. 3, pp. 399–423.

Sassoon, D. (1981) *The Strategy of the Italian Communist Party* (London: Frances Pinter).

Scott, D. (1994) *Formations of Ritual* (Minneapolis: University of Minnesota Press).

Scott, J. C. (1985) *Weapons of the Weak* (New Haven: Yale University Press).

Sereni, E. (1947) *Il Capitalismo nelle Campagne* (Turin: Einaudi).

Sewell, W. H., Jr (1990) 'How classes are made: critical reflections on E. P. Thompsons's theory of working-class formation', in Kaye, H. J. and McClelland, K. (eds) *E. P. Thompson. Critical Perspectives* (Philadelphia: Temple University Press) pp. 50–77.

Shore, C. (1990) *Italian Communism: The Escape from Leninism* (London: Pluto).

Silber, L. and Little, A. (1995) *The Death of Yugoslavia* (London: Penguin Books).

Silverman, S. (1965) 'Patronage and community-nation relationships in central Italy', *Ethnology*, vol. 4, pp. 172–89.

— (1970) 'Exploitation in rural central Italy: structure and ideology in stratification study', *Comparative Studies of Society and History*, vol. 12, pp. 327–39.

— (1975) *Three Bells of Civilization* (New York: Columbia University Press).

— (1977) 'The myth of patronage', in Gellner, E. and Waterbury, J. (eds) *Patrons and Clients in Mediterranean Societies* (London: Duckworth) pp. 7–20.

Simic, A. (1973) *The Peasant Urbanites* (New York: Seminar Press).

Smith, A. D. (1995) 'Gastronomy or geology? The role of nationalism in the reconstruction of nations', *Nations and Nationalism*, vol. 1, no. 1, pp. 3–23.

Smith, G. (1991) 'Writing for real', *Critique of Anthropology*, vol. 11, no. 3, pp. 213–32.

Snowden, F. M. (1972) 'On the social origins of agrarian fascism in Italy', *European Journal of Sociology*, vol. 13, pp. 268–95.

— (1979) 'From share-cropper to proletarian: the background to Fascism in rural Tuscany', in Davis, J. A. (ed.) *Gramsci and Italy's Passive Revolution* (London: Croom Helm) pp. 136–71.

— (1986) *Violence and Great Estates in the South of Italy* (Cambridge: Cambridge University Press).

Sorabji, C. (1989) Muslim Identity and Islamic Faith in Sarajevo (Cambridge: University of Cambridge) unpublished Ph.D thesis.

— (1993) 'Ethnic war in Bosnia', *Radical Philosophy*, vol. 63, pp. 33–5.

— (1995) 'A very modern war: terror and territory in Bosnia-Hercegovina', in Hinde, R. A. and Watson, H. E. (eds) *War, Cruel Necessity? The Bases of Institutionalized Violence* (London: I.B. Tauris) pp. 80–95.

Spriano, P. (1975) *The Occupation of the Factories: Italy 1920* (London: Pluto Press) translated by G. Williams (ed.).

Sullivan, J. (1988) *ETA and Basque Nationalism: The Fight for Euskadi 1890–1986* (London: Routledge).

Taggart, P. (1995) 'New populist parties in Western Europe', *West European Politics*, vol. 18, no. 1, pp. 34–51.

Therborn, G. (1995) *European Modernity and Beyond* (London: Sage).

Thompson, E. P. (1963) *The Making of the English Working Class* (London: Penguin).

Tilly, C. (1978) *From Mobilization to Revolution* (New York: Random House).

Touraine, A. (1981) *The Voice and the Eye* (Cambridge: Cambridge University Press).

— (1985) 'Sociological intervention and the internal dynamics of the Occitanist movement', in Tiryakian, E. A. and Rogowski, R. (eds) *New Nationalisms of the Developed West* (Boston: Allen and Unwin) pp. 157–75.

Touraine, A. and Dubet, F. (1981) *Le Pays Contre L'Etat* (Paris: Seuil).

Urbinati, N. (1998) 'The Souths of Antonio Gramsci and the concept of hegemony', in Schneider, J. (ed.) *Italy's 'Southern Question': Orientalism in One Country* (Oxford: Berg) pp. 135–56.

Urla, J. (1993) 'Contesting modernities: language standardization and the production of ancient/modern Basque culture', *Critique of Anthropology*, vol. 13, no. 2, pp. 101–18.

van de Port, M. (1999) 'It takes a Serb to know a Serb', *Critique of Anthropology*, vol. 19, no. 1, pp. 7–30.

Vasic, M. (1996) 'The Yugoslav army and the post-Yugoslav armies', in Dyker, D. and Vejvoda, I. (eds) *Yugoslavia and After* (London: Longman) pp. 116–37.

Vejvoda, I. (1996a) 'Yugoslavia 1945–91: from decentralisation without democracy to dissolution', in Dyker, D. and Vejvoda, I. (eds) *Yugoslavia and After* (London: Longman) pp. 9–27.

— (1996b) 'By way of conclusion: to avoid the extremes of suffering', in Dyker, D. and Vejvoda, I. (eds) *Yugoslavia and After* (London: Longman) pp. 248–63.

Verdery, K. (1994) 'Ethnicity, nationalism and state-making', in Vermeulen, H. and Govers, C. (eds) *The Anthropology of Ethnicity* (Amsterdam: Het Spinhuis) pp. 33–58.

Visentini, T. (1993) *La Lega: Italia a Pezzi?* (Bolzano: Edition Raetia).

Weine, S. M. (2000) 'Redefining Merhamet after a historical nightmare', in Halpern, J. M. and Kideckel, D. A. (eds) *Neighbors at War* (Pennsylvania: Pennsylvania State University Press) pp. 401–12.

Wicker, H.-R. (1997) 'Theorizing ethnicity and nationalism', in Wicker, H.-R. (ed.) *Rethinking Ethnicity and Nationalism* (Oxford: Berg) pp. 1–42.

Wiegandt, E. (1993) 'Towards an understanding of rural ethnic mobilisation 1850–1940', in Howell, D. (ed.) *Roots of Rural Ethnic Mobilisation* (Dartmouth: New York University Press) pp. 305–19.

Wieviorka, M. (1993) *The Making of Terrorism* (Chicago: The University of Chicago Press).

Williams, G. (1975) *Proletarian Order* (London: Pluto Press).

Williams, R. (1985) *The Country and the City* (London: Hogarth Press).

Willson, P. (1993) *The Clockwork Factory. Women and Work in Fascist Italy* (Oxford: Clarendon Press).

Wood, E. M. (1986) *The Retreat From Class. A New 'True' Socialism* (London: Verso).

— (1995) *Democracy Against Capitalism* (Cambridge: Cambridge University Press).

Woodward, S. (1995) *Balkan Tragedy. Chaos and Dissolution after the Cold-War* (Washington, DC: Brookings Institute Press).

— (1996) 'The West and the international organizations', in Dyker, D. and Vejvoda, I. (eds) *Yugoslavia and After* (London: Longman) pp. 155–76.

Zulaika, J. (1988) *Basque Violence: Metaphor and Sacrament* (Reno and Las Vegas: University of Nevada Press).

Zulaika, J. and Douglass, W. A. (1996) *Terror and Taboo* (London: Routledge).

INDEX

Compiled by Sue Carlton

Printed and bound by CPI Group (UK) Ltd, Croydon, CR0 4YY

27/10/2024

14580225-0002